BUREAUCRACY AND THE NEW AGRICULTURAL STRATEGY

BUREAUCRACY AND THE NEW AGRICULTURAL STRATEGY

KULDEEP MATHUR

CONCEPT PUBLISHING COMPANY
NEW DELHI

First Published 1982

Kuldeep **Mathur** (*b*. 1938-)

Published by
Naurang Rai
Concept Publishing Company
H-13, Bali Nagar
NEW DELHI - 110015 (India)

Printed by
Mohan Printing Corporation
Padam Nagar
Delhi - 110007

ACKNOWLEDGEMENTS

THE PRESENT study is part of a three nation study sponsored by the Asian Association of Social Science Research Councils (AASREC) Bangkok. It was financed through a grant from the UNESCO office of the Social Science Regional Adviser, Bangkok. The preliminary draft of the Report was discussed at a Seminar convened by AASREC and the Regional UNESCO office Bangkok. In finalizing this study, suggestions made during the Seminar and by Dr. Yogesh Atal, the Regional Adviser, were very helpful.

I am also greateful to Shri T.N. Chaturvedi, the then Director of the Institute for providing facilities for carrying out this work. Dr. T.S. Yadav provided useful research assistance and carried out most of the field work. Shri Anil K. Gupta read the earlier drafts; his critical remarks have particularly helped in improving data analysis and presentation. Prof. Harry J. Friedman provided opportunities for discussion on several substantive issues and helped in sharpening their focus.

Finally, I would like to express my deep appreciation of the diligent and efficient typing of the script that was undertaken usually at short notice by Shri S. Prakash and Shri M.P. Gupta.

KULDEEP MATHUR

LIST OF FIGURES

CONTENTS

LIST OF TABLES

1

Introduction

SINCE the advent of the planning era, considerable interest has been evinced in investigating the role and working of bureaucracy in India. The reason for this interest is not far to seek. A former colonial nation was attempting to transform itself socially and economically and had placed heavy responsibility on its own bureaucracy to do so. So the important question that was raised was whether the bureaucracy inherited from the colonial days was capable to accept this challenge or not. Most observers doubted this capability (Appleby, 1953; 14, La Palombara, 1963; 14, Braibanti, 1966) and argued more or less that "... the inadequacies of the Indian bureaucracy are not due to the fact that it is a bureaucracy but due to a considerable extent to the fact that it carries too much baggage from the past". (Taylor *et. al.* 1966 : 579).

Nehru, voiced a similar sentiment when he pointed out that "... no new order can be built up in India so long as the spirit of the ICS pervades our administration and public services". (Nehru, 1953 : 282). He and other political leaders joined in giving exhortations to the bureaucracy to change their orientations to serve the people and drop their colonial frame of reference. The following dictum of Mahatama Gandhi was often quoted :

> Whenever you are in doubt or the self becomes too much with you, apply the following test. Recall the face of the poorest and weakest man whom you have seen and ask yourself if the step you contemplate is going to be of any use to him.

The crucial question that was posed was whether the traditional bureaucratic structural and behavioural norms were at all compatible with new task of administering development. Much of the empirical work on Indian bureaucracy dealt with such an issue (e.g. Mathur, 1972a, 1972b; Pai Panandiker and Kashirsagar, 1978; Bhambhri, 1972) and attempted to present profiles of behavioural orientations in order to examine how they did not match with the expected profiles of developmental administrators. The basis of such studies was that the performance of the administrators was dependent on their mental make up and a 'developmentalist' was needed to implement development policies.

With the advent of community development and Panchayati Raj, the developmental scene tended to highlight the relations between political leaders and administrators as being crucial in implementing change in the rural areas. It was argued that the success of local institutions as instruments of development rested on the ability of the political and administrative leadership to cooperate with one another and resolve the tensions and conflicts arising from the social environment. The empirical researchers who took this direction investigated the areas of conflict and analysed the reasons of the emergence of tensions. (Kothari and Roy, 1969; Gaikwad, 1969; Bjorkman, 1976; Chaturvedi, 1964). The two types of studies on bureaucracy and implementation processes mentioned above looked at bureaucratic behaviour as a product of inter-personal relations and group dynamics occurring within specific organizational boundaries.

In addition, most of the researches cited above based their argument on an instrumental view of bureaucracy. In this view, bureaucratic and political processes were considered separate and it was assumed that the role of bureaucracy was to implement policies as effectively as possible. The success of developmental plans depended on how well bureaucracy implemented them. Implicit in these assumptions was that the plans were well prepared and well-conceived and if the country could not achieve the goals and targets set, it was due to faulty implementation. Strengthening the attributes required of the so called development administrator was considered as a necessary ingredient for improving bureaucratic performance and for achieving development targets.

Doubts began to be raised about this instrumental role when a call for committed bureaucracy was made by the Prime Minister in early 1970s. The probable thrust of the argument was that bureaucracy was as much part of the political processes and the way plans were formulated and implemented depended on a great deal on how the bureaucracy perceived problems and identified methods to solve them. Such a perception was dependent on bureaucracy's own understanding of its place in society and its views on changes sought to be brought about. It was possible for bureaucracy to choose its own lines of action in conformity with its own interests and these may or may not conform to the ones intended by the development plans or the policy makers. Alternatively, bureaucracy and policy makers could combine and forge identity of interests and then work in favour of a given policy or against it. In this sense bureaucracy was seen as no different from the political groups who formulate policies and it began to be advocated that implementation is tied organically to the policy planning processes.

One of the implications of this line of reasoning is that the role of bureaucracy and its working is intimately connected with the way policies are framed and what the 'real' intentions of the formulators of policies are. Bureaucracy carries little capacity to function autonomously and perform tasks that are not envisaged in the 'will' embedded in the policies. Thus, exhorting the bureaucracy to work for the rural poor without providing the necessary policy and implementation framework is either only a political rhetoric or an obvious contradiction of demands placed on it. The case of implementation of land reform policies is a clear illustration of such dynamics at work. (cf. Mathur, 1980).

Thus, over the course of years, understanding of the role of bureaucracy in implementing development plans has undergone a change. This change has come together with the emergence of a new perspective on the role of public administration itself in development. After the Second World War and with the advent of the philosophy of foreign aid to trigger off development, it was assumed that once decisions were taken and policy objectives were defined, public administration could translate them into action because it was further assumed that technical solutions were available to public problems. The recognition of the intimate relationship of the phenomenon of administration to the public realm was submerged in favour of an emphasis on administration as a generic process susceptible to scientific study if freed from the trammels of political value or moral dimensions. (Savage, 1974 : 150), Public administration was seen as any other profession that increased its ability by gaining greater technical competence.

As the developmental scene unfolded itself in the Third World, it became more and more difficult to view public administration in the above perspective. Administration itself became a dimension of development and it began to be recognised that administration was embedded intimately in the socio-political context of development and its autonomy was limited, to say the least. Institutions of public administration were now seen to represent the social forces and interests dominating at a particular period of time. What emerged was a far more dynamic view of the role of administration in development. Consequently, administration was development process in action. And if development is propelled by classes and groups interested in a new social and economic order, it will be obstructed by those who would like to see the preservation of the status quo. Administration would reflect such pulls and pressures and would not be as autonomous of them as viewed earlier.

Views about the role of bureaucracy have reflected these changes. Its effective functioning is not dependent on its internal competence alone but is determined to a great extent by the policies, logic and intentions it is supposed to implement. Some have argued that bureaucracy is a deliberately designed social structure which has been established by a certain person, group or class in order to implement certain goals (Abrahamson, 1977 : 11), Marxist tradition treats bureaucracy as one element in the power exertion of the ruling class. What this tradition emphasizes is that bureaucracy cannot be treated as an independent unit of analysis. It must be scrutinized against the social and economic background of those who rule and its capacity to implement programmes analysed in light of the real intentions of policies and decisions formulated.

The purpose of this brief study is embedded in the broad focus enumerated above. The study seeks to examine the way bureaucracy performs its role and functions within the context of national policies and politics and administrative framework. It seeks to develop hypotheses to argue that bureaucracy has little role to bring about change in society unless policy and administrative framework transform themselves concomitantly to support bureaucratic action. The proposition is that changes in role and behaviour of Indian bureaucracy are expected when the essential structure and norms legitimizing them do not change or are not changed deliberately. Minor adjustments are made, to be sure, but these are not enough to sustain new outlook, new behavioural patterns and new methods of working.

An attempt has also been made to relate the problems of local level administration to those at higher levels of administration. It is only through the investigation of this relationship that the contours of constraints in implementation can be discerned. In the realm of rural development administration this inability to look at these linkages appears to be quite marked.

Thus, in this study an attempt is made to place the role of bureaucracy in the wider framework of policy implementation processes. Policy implementation process is a continuous decision-making activity in which conscious choices about strategies of implementation are made. Policy goals, programme choices and implementation strategies are linked together in producing policy or programme outputs. The kind of output and its level are determined by the interaction of all these components of the implementation process. To assume that only one component is the determinant leads to the creation of artificial autonomy that is not warranted by the empirical processes of implementation.

In addition, the implementation process tends to include the social and political processes that influence its course. It accepts that what is implemented may be the result of political calculus of interests and groups competing for scarce resources. Looking at functioning of bureaucracy within this framework may imply an assessment of political and power capabilities of groups and interests influencing it. More often than not, administration in India is the arena where most political battles are fought. Including these battles within the purview of this study will lead to better understanding of the role of bureaucracy.

The study has been conducted in the context of the implementation of Green Revolution policies in a district Karnal located in Haryana, a State in North West India. In view of the focus of inquiry described above, there are two distinct parts of the study. In the first part the national policies towards rural development and the national administrative framework has been presented and analysed. In the second part, empirical data on the actual processes of implementation in the district have been presented. A sample of blocks and villages and respondents of officials, political leaders and cultivators was chosen for this empirical study and the process of selection is described later. Finally, an attempt at synthesizing the two parts is made. Both the primary and secondary data were collected during the later half of 1979.

2

The Administrative Framework : Federation, Political Processes and the Bureaucracy

IN UNDERSTANDING the role of bureaucracy in implementing tasks of development, such features of the constitutional framework, that influence it need to be delineated. These features determine, to a great extent the structure of bureaucracy in India and also define the competence of levels of governments and the functions allotted to it. Constitutionally, India has a federal form of government wherein conscious effort has been made to define the area of policy formulation and implementation between the Centre and the States. Clear listing of powers has been made through the Central List, State List and the Concurrent List in the Constitution. The way the Federal system actually works, however, not only depends on the directions given by the Supreme Court in interpreting the Federal provisions but also on the bargains struck by the political forces operating in the country at a given period of time.

The Union has enjoyed wide powers for formulating policies in the general area of economic and social planning and in the individual sectors of the economy. In the particular case of rural development, innovative policies has been articulated by the Centre even though 'Agriculture and Rural Development' is a state subject. It has usually prepared guidelines for the strategy of change and determined the pattern of programmes that will implement it. In many cases, the implementation of programmes involves both the Centre and the States, and the two jointly share the responsibility for results. The Community Development Programme is one example of the frequently observed arrangements wherein the states have enacted their specific laws to usher in new institutional framework while the Centre has not only determined the institutional framework but also kept initial control and regulation of the programme through grant of funds. Total management of these agencies is taken over completely by the States only when the Centre so desires.

The result of such an arrangement is that to sustain a scheme, the Centre and the States must keep up their interests in it. The scheme can be maintained by Centre's interest if it continues to give financial grants. So long as the resources of the Centre are forthcoming, it is for the benefit of the States to keep up its interest in form at least if not in substance. This means that, more often than not, the involvement of the States may be determined by the quantum of funds it is going to receive from the Centre and not necessarily by the acceptance of objectives of rural development by its leadership. Consequently, states give great importance to the bureaucratic paraphernalia that has to be installed in order to see that funds keep flowing in and often ignore the objectives for which it has been set up.

This leads to a strange situation. The Centre is free to articulate a specific policy and allocate funds for its implementation and the states, on the other hand, accept the funds without being committed to fulfil the objectives of the policy.

Institutional arrangements exist to involve both the Centre and the States in policy formulation. For purposes of determining policies a National Development Council has been set-up. The First Five Year Plan said : "in a country of the size of India where the States have under their constitution, full autonomy within their sphere of duties, it is necessary to have a forum such as National Development Council in which from time to time, the Prime Minister of India and Chief Ministers of States can review the working of the Plan and its various aspects". What transpires in the Council meetings depends upon the prevailing political balance in the country. Frequently, the National Development Council has been used by the States to ventilate their differences with the Centre and utilize the forum to arrive at compromises.

Thus, the federal system has created unique problems. On the one hand, what the Central Government can or cannot do in effecting socio-economic changes in the rural areas depends to a great extent on what the States want or do not want to do. The states have their own pattern of power distribution and they also command the administrative machinery and therefore, their potential to disregard central dictates exists. On the other hand, as a few States not ruled by the same party as at the Centre, have argued their capacity to bring about socio-economic changes depends to a great extent on whether the Central financial dispensations are adequate or not. It is clear, therefore, in the field of rural and agricultural development there is a strong link between the Centre and the States. The strength and directions of this link is not only affected by the formal division of powers but also by the political equations prevailing in the country.

The balance of power between the Centre and the States shifts as these political equations change. But the relations are critical to India's development. As the intermediary between the worlds of abstract planning and village reality, the State Governments are strategically placed to abet or retard the development process. (Viet, 1976, 52). The character of relations depends on the extent of inter-state rivalry, policies and demands of the states, actual and perceived high-handedness of the Centre and the nature and background of the political elite at both levels.

Administration and bureaucracy

The administrative system follows the pattern of the federal structure. Both the Centre and the States have their separate administrations governed by independent sets of personnel policies. The state administrative structure pervades through the district to the village level. Thus, the machinery that usually implements development plans belongs to the State Government. The central administration chooses to provide guidelines, policy framework etc. for implementing specific policies. Increasingly in the past decade, the Central Government has been intervening at the local level through what have come to be known as centrally sponsored schemes in which it determines the organizational design, financial outlay, goals and targets etc. But even in these cases it has to depend on state personnel who staff these organizations.

But the very nature of tasks allotted to the States, functional departments implementing developmental plans in the rural areas are with the State Government. Similar departments with large bureaucracies have emerged at the Centre too but their tasks are essentially in the area of policy planning, monitoring, evaluation etc., or what are known as staff functions.

The impact of the first wide scale rural development progrmme of Community Development was an unprecedented growth of State departments and bureaucracy. Functional organizations with their hierarchies reaching out from the State to the district, block or even village level multiplied. The districts of the country were divided into blocks which formed the basis of community development programme. An extensive network of village level workers, extension officers and block development officers was built up. It is this network that has formed the core of the bureaucratic system in the rural areas and has survived most diversification and expansion. The following Table 2.1 provides details of this network.

TABLE 2.1 : **Extension staff position—1977**

	No. of workers	Per thousand families	No. of families/ workers
B.D.O.	4,777	.06	17,668
Gram Sewak (VLW)	65,450	.78	1,290
E.O. (Agriculture)	6,096	.07	13,845
E.O. (Animal Hus.)	5,046	.06	16,726
E.O. (Industries)	695	.01	1,21,439
E.O. (Cooperatives)	4,208	.05	20,057
E.O. (Edn./SE. Org.)	1,810	.02	46,630
E.O. Panchayat	4,435	.05	19,030
Gram Sevika	6,004	.07	14,057
Mukhya Sevika	1,527	.02	55,272
Overseer	4,961	.06	17,030
Medical Officer	7,717	.09	10,937

Source : Government of India, Ministry of Agriculture, 1978-79
BDO=Block Development Officer
E.O.=Extension Officer
Gram Sevika=Lady Village Level Worker

The local bureaucracy so created was connected to the state administrative system which responded by creating and expanding its own generalists and technical cadres. Thus, what happened was that multiple bureaucratic cadres emerged and each had their links with the village or block or district levels. A variegated bureaucratic network with long lines of hierarchy gradually began to unfold itself. The Green Revolution strategy relying primarily on technological innovation considerably influenced the growth of technical cadres. For the supply of each of the inputs into agriculture planning, a new organization and hence a new cadre grew up.

The IAS is the elite corps of Indian Administration and holds key secretariat and field positions both at the Central as well as the State levels. Members of this Service are also District Collectors/Magistrates and thus the key functionaries in the District. While many organizations and cadres have been added to the district, the District Collector has not been disturbed in his role as the primary officer in the district. In most cases, he heads the new organizations too and both the State and Central Governments place responsibility on him for all developmental or regulatory work in the district.

The characteristics of the structure of the Indian Administrative Service (IAS) reflect the way the federal system works. There is a duality of control over the IAS with the Central Government acting as a check over the State Government. The recruitment is done by the constitutionally established Union Public Service Commission and induction training is provided by a Central Government Institution. When a member of the IAS is confirmed into service he is allotted into the cadre of any one State Government. But during his career, he may also serve the Central Government. When he does so he is placed on deputation to it usually for not more than five years after which he reverts to his own state cadre. During his service with the State Government, however, he is governed by rules and regulations framed by the Central Government even though his postings and transfers are decided by the State Government. Consequently in case a member of the IAS feels that the State Government has altered or interpreted such rules and regulations to his disadvantage, he can appeal to the Central Government. This is supposed to guarantee his independence and assures him immunity from the state political processes. This arrangement in which the top positions in the State bureaucracy are held by administrators who owe loyalty to the Centre and shuttle between the Centre and States further ensures that the states reflect the central will in matters of rural and agricultural development which is constitutionally a state subject.

The fact that the members of the IAS hold crucial positions in Government at all levels and that this Service traces its lineage to the Indian Civil Service (ICS) of the pre-independence era makes the IAS very influential in the governmental machinery. It provides leadership in all tasks and sets examples to other cadres by its attitudes, values and behaviour. For this reason, when the question of compatibility of the inherited administrative machinery with the challenge of development tasks was raised much of the attention was focused on the IAS. The Five Year Plans exhorted the bureaucracy to change itself by inculcating a spirit of dedication and help towards the rural poor. (see Chapter 1).

However, the structure of the service has remained in tact since the last three decades. There have been some changes, though. The last of the old ICS retired in 1980 and the IAS men who replaced them hold the more powerful positions in government to-day. The so-called 'steel frame' of the earlier days has rusted to the extent that new cadres have made in roads into its dominating position. In addition, the annual in take, on the average, has been of more than hundred persons thus reducing the exclusiveness that the ICS enjoyed in the colonial days. For these reasons, there has been some dispute whether such a legacy still continues to pervade the Indian administrative system or not. Drastic structural changes have not occurred and levels of performance have continued to receive critical attention throughout the planning decades.

Political processes and rural scene

During the period of independence struggle, the national leadership had committed itself to rapid rural development through drastic changes in pattern of land holdings and social relations. The commitment stemmed from the fact that the independence was spearheaded by the urban elite and the rural elite consisting of big landlords or zamindars were clearly identified with the British rulers. Therefore, initially the whole attempt at the abolition of intermediaries was aimed at the zamindari-jagirdari system and the end of the British rule logically had to be followed by the elimination of its allies. The relative success achieved in this regard was really another facet of the achievement of independence itself.

Once this was done, it was difficult for the central leadership to move ahead. The commitment started waning for several reasons. Among them, an important one was that the character of leadership started changing at the central level. The occupational background of the members of the lower house of the national parliament, Lok Sabha, shows a shift in pattern from 1951 to 1977. The First Lok Sabha had 19 per cent of its members who acknowledged agriculture as their occupation. Over the years, this proportion gradually increased to 38 per cent in the Sixth Lok Sabha. Business and law which was represented by 35 per cent of members in 1951 shrank to 25 per cent in 1967 and stabilized to 23 per cent in 1977. Members of the agricultural occupation have continued to increase their representation with each election and form the single largest group in the Sixth Lok Sabha.

The rise of the influence of the agricultural classes at the Central level was buttressed by concomitant growth of the power of the rural landed interests. Two significant government policies among many others aided this process. In 1952, the Government of India introduced community development schemes all over the country. In 1959, Panchayati Raj was introduced to support community development through local participative organizations.

The major contribution of Panchayati Raj was to introduce elections with the possibility of gaining access to decision-making bodies right from the village to the district. This was also the area where traditional caste leaders dominated. Now an opportunity was shown for those who could build alternate sources of power to compete with them. Economic base was the only alternative source of power. Consequently, land as source of power emerged as a strong factor in local politics. With it emerged those leaders who knew that any effort to deprive them of land would mean drastic reduction in the power they wield. Thus, as the Panchayati Raj institutions took hold, the power of the land owning classes increased and stabilized itself.

Looking at this dynamics from the point of view of state and national politics, even more serious repercussions have been the links that this leadership forged with each other. The members of state assemblies and Parliament quickly realized that to gain rural votes, it was important to win over Panchayati leaders. They controlled the rural votes and could sway the population in favour of one person or another. Panchayati Raj became a vote bank for political purposes. The result was that leaders tied to local issues commanded greater influence than hitherto and because of the political linkage formed during the elections, they had the ability to ask for *quid pro quo* from those at the National and state level who had arrived because of their support. Because of closeness to local issues, state legislatures are characterized by much greater representatives from the richer agricultural classes. In the federal set up, it is these legislatures that enact legislations to implement policies for rural development.

The second series of factors that helped in the rise of the influence of the richer agricultural classes are concerned with policies towards land reform and introduction of new technology in agriculture. When the intermediaries were abolished and about 20 million cultivators were brought into direct contact with the state, compensation was promised to those who were divested of land. The total compensation including rehabilitation grant and interest was estimated in 1969 to be Rs. 604.1 crores out of which Rs. 281.10 crores had so far been paid to the ex-intermediaries either in cash or in the form of bonds. (*Yojana*, 1969 : 5).

With this capital, the land owning groups diversified their interests and invested in commercial activities. In addition, this capital gave them the capability to modernize agriculture and invest in improvements in land. It allowed them to use land reform measures to evict tenants, employ labour and claim more and more land for personal cultivation.

This process was hastened with the introduction of new technology that made agriculture a profitable proposition. Only those farmers could use modern inputs and techniques who had the requisite financial resources to invest. Obviously, the capacity of the large farmer to invest or borrow money was much greater than that of the small one. The governmental programmes to strengthen the small farmer through loans and grants did not work adequately because the politically powerful had the ability to direct such efforts in their direction. Thus, the new agricultural developments took place in the farms and fields of rich and larger farmers. Consequently, the richer agricultural classes have also become symbols of agricultural transformation and have impeded effectively such land reforms measures that seek greater equality in land distribution in the name of progress and modernization of agriculture.

Thus, the rising influence of the agricultural classes is a striking feature of the dynamic political processes of the past six elections. Rajni Kothari writing in 1970 said : "the newest power group in Indian politics is that of the Kulaks, the class of independent owner-cultivators drawn from a variety of social groups who are making agriculture a thriving business proposition and one that provides them with a base for effective political bargaining at higher level", (Kothari, 1970 : 354). The current situation in the country is reminiscent of the historical dualism between village power and central power in which cadres of administrators play the intermediary role, (Veit, 1976 : 28). Rural urban tensions are on the increase and the 1979-80, Union Budget has been characterised as a 'rural' budget attempting to focus attention on the problems of the farmers. Another manifestation of these tensions is the demands of the new rich agriculture classes for places in the bureaucracy and urban professions. However, they are at a disadvantage in this respect with the urban elite who have a better command of English and are able to utilize city education facilities more fully. Consequently, the policies of state and central governments allowing job reservations and use of regional languages for competitive examinations is designed to safeguard the interests of the rural elite.

Within the rural sector, the attempt at terrorising the poor reflects the concern of the elite for maintaining economic power through assured supply of cheap labour. Recently, violent demonstrations of the richer classes of a village took place in Delhi to ask the government to evict the scheduled castes (usually the agricultural labourer and the landless) from the lands that had been allotted to them in 1970. Numerous other incidents of village atrocities are being reported throughout the country.

Finally, one needs to emphasise that the major cause of the present build up on tensions is also that the same factors that have led to the prosperity of the well-to-do farmers by commercialising agriculture and rationalizing the processes of production have also contributed to greater awareness of prospects of economic improvement among the Harijan marginal peasants and landless agricultural labourers. Obviously, this means that the poor are now more disposed to resist the kind of exploitation and indignities which they had earlier put up with.

Concluding remarks

At the local field level where programmes are implemented consequences of the above features of Indian polity are many. The administrative institutions that exist reflect the federal situation. There are schemes that are State Schemes and are implemented by the State bureaucratic hierarchy. This hierarchy is not one but fragmented into many sub-hierarchies. Thus, each sector will have its own bureaucracy governed by its own rules and regulations and career opportunities. Then, there are centrally sponsored schemes which are termed as Special Schemes which are financed by the Centre and implemented through newly created agencies. Guidelines for formulation and implementation of projects under these schemes are determined at the Central level which further determines their financial allocations. State Governments have little say in the affairs of these schemes though they depute their own officers to the agencies that implement them. There can also be purely Central projects that are entirely implemented by the centre providing finance, personnel, etc. There are also local projects implemented by local elective bodies like the Panchayats.

Such diverse administrative agencies are now linked with the political agencies. We have already noted that due to variety of reasons the power of agricultural classes has increased over the years. When they are unable to influence the formulation of policies they shift their interest to the arena where these policies will be implemented. Using variety of influences, vested interests seek to give direction to implementation to ensure for themselves benefits. With links already forged with the Centre and State political leadership, local political leaders attempt advantages through relocation of schemes or forcing leakages in distribution of scarce resources etc. A web of political and administrative relationships is spun and changes its patterns to reflect the flow of advantages.

Further, a dominant orientation of the administration is that of law and order. The District Collector considered internal peace crucial to good administration in his area. His esteem among his peers depended largely on how he performed his duties in this sphere. Lack of initiative or lack of judgment in face of an unruly mob or failure to exercise authority to quell a riot were the most unwanted qualities that an officer could possess.

A serious consequence of such orientation is that when the share-cropper or the landless mobilize support to fight for their rights, the resulting confrontation between them and the landlords is seen as breach of peace to be dealt with routine law and order instruments. Thus, the adminstrative system becomes status-quo oriented and oppressive towards those who demand change. Therefore, a dedicated administrator here and there who supports change faces insurmountable difficulties from both his political as well as administrative bosses. Within this orientation, the District Collector finds himself burdened with roles that sometimes contradict each other. Usually, the contradiction is resolved in favour of law and order role over developmental functions.

The emergence of dominance of law and order role means that in a society where conflicts are emerging due to developmental politics, bureaucracy becomes an ally of the dominant group in the power structure. This is in its own interest of protecting itself from the on slaughts of those classes who are a serious threat to its existing role and functioning. Consequently, an important response of the bureaucracy to the situation of conflict is to insulate itself through myths of neutrality and responsibility to the Constitution alone and not to any political party, (Mathur, 1971). Such a response drives a further wedge between the bureaucracy and the public on one hand and bureaucracy and its political masters on the other.

A major problem that emerges from this discussion is that the bureaucracy has not been able to enlist support for its activity precisely from those groups which it was intended to serve. As far back as 1961, the Third Plan had pointed out that the administration has failed to enlist "the support and sanction of the village community in favour of effective enforcement of legal provisions". The dilemma facing the country is whether such support can be enlisted without radically transforming the role and functioning of the bureaucracy as it obtains today.

3

National Context of Policies Towards Rural Development

BROAD economic objectives of growth, distributive justice and self-reliance have been widely accepted as the cornerstones of India's planning for development. However, there has been a continuous debate about the strategy of achieving these objectives. The argument has been based on the question of choice of emphasis on agricultural or industrial sector. To many, agriculture has been neglected in Indian planning. Invoking Gandhi, such people feel that western industrialization or technological progress is not an answer to India's problem. Solutions have to be found in the direction of regenerating Indian village communities through small scale agriculture and village industries. This involves minimum use of machinery and maximum use of labour. Gandhi had rejected class conflict and conceived of a society where injustices could be removed peacefully through moral force and village would emerge as a place of peace and harmony. Voluntary change of heart should be used as a major lever of change.

Nehru, on the other hand, was the prime advocate of Indian socialism which owes its origin to Democratic Fabianism and Soviet Marxism. While the former stressed social ownership of means of production and egalitarian distribution of output, the latter demonstrated the speed at which modernization of economy can be done. Both of these ideas were attractive to the Indian national elite and were enshrined in the developmental strategy. Instead of personal voluntary action, purposive state action was relied upon to bring about change. Thus, while Gandhism continued to be given respect for its higher ideals and moral exhortations, the Plan documents gave place to a variant of the Harrod-Domar model which distinguished two major sectors—capital goods and consumer goods—and went on to use the industrial sector as a vector of development. The planning effort also recognized public sector as an important means for achieving developmental objectives and sought to raise it to "commanding heights" of Indian economy. Finally, within the agricultural sector, the planning effort was to facilitate agricultural production through peasant proprietorship by providing inputs, infrastructural facilities and institutional reform.

The result of such a strategy has been that, by and large, a dependent and stagnant economy has become modernized and made self-reliant in certain sectors. Agricultural output has risen throughout this period, more so after the advent of Green Revolution. Industrialization has also progressed at a pace which has made India self-sufficient in most consumer goods and capable of producing capital and basic goods like steel, fertilizer etc.

However, as the Draft Plan (1978-83) points out, " . . . the pace of industrialization has not been bought cheaply. The concentration of economic power has increased, in the sense that within the corporate sector, the assets of bigger corporations have increased more rapidly. The expansion of large scale industries has failed to absorb a significant proportion of the increment to the labour force . . ." (p. 2). In the area of agriculture, growth is spread unevenly. Per capita agricultural production has remained stagnant. In addition, the Draft Plan (1978-83) emphasizes that "the percentage of population below the poverty line in 1977-78 may be projected at 46 per cent in rural areas". (p. 6).

Thus, the development scene as a consequence of the strategy followed in India is marked by severe disparities in the distribution of benefits of growth. Islands of prosperity in a vast ocean of poverty exist and the Draft Plan (1978-83), itself is constrained to remark that "we must face the fact that most important objectives of planning have not been achieved, the most cherished goals seem to be almost as distant today as when we set out on the road of planned development". (p. 4)

Rural development : Policies and strategies

Within this overall perspective, policies towards rural development have taken several directions. Efforts have been made to change the structure of rural community through land reform, facilitate new institutional framework through community development and Panchayati Raj, provide infrastructural facilities for increasing agricultural production directly or through expansion of credit facilities, and establish new institutions to promote the interests of a specific area or a particular target group. Usually, there are multiple policies working at the same time but receiving different emphases. Rarely, has a particular programme been replaced by another. More often than not, old programmes are allowed to languish when new ones are taken up.

Land reform

Land reform is crucial to rural development and was among the policies that found place even in the pre-independence manifestoes of the Congress Party. In most rural areas of the Third World, poverty is intimately related to the degree land is concentrated, (ILO, 1978 : 34). It is due to this reason that it has been argued that "rural development cannot be said to have begun without land reform" (Hagne, *et al*, 1975 : 96).

Land reform policies in India have been concerned with action on three fronts : (i) the abolition of intermediary tenures like zamindaries, jagirdaries, and inams which covered 40 per cent of the area of the country, (ii) providing security of tenure to tenants and regulation of rents, (iii) the imposition of ceilling on land holdings and the distribution of surplus land to the landless and uneconomic landholders.

It is in respect of the abolition of intermediary tenures that the success has been almost complete. More than 20 million cultivators have been brought in direct relation with the State. Policies in the area of tenancy legislation have had uneven implementation. Tenants, by and large, do not enjoy fixity of tenure still. They are either tenants at will or subject to landlords rights of resumption or enjoy temporary protection only. There has been leasing of land on a considerable scale even in areas where intermediary tenures existed. Such tenancies are generally of an informal nature without written document or record which makes it difficult to arrive at a correct assessment of the extent of the tenancy problem. Provisions regarding fair rent have largely remained unimplemented and the general pattern is of demanding and paying fifty per cent of the produce as rent, (Appu, 1975).

Policies for achieving greater equality in the agrarian structure were directed towards determining ceilings on land holding and redistributing surplus land made so available among the landless. The 1954-55 data show that 85 per cent of the small and medium land holding households (up to 4 ha.) accounted for a little over one-third of the total cultivated area. At the other end of the scale, the top 15 per cent of the households accounted for about two-thirds of the total area. The 1971-72 data showed that though the number of households owning holdings above 4 ha. registered a slight decline, more significant change was noticed in the area cultivated by these holdings which came down to little over 50 per cent. However, as one can see from the Table 3.1 below, the pattern of unequal distribution of land has maintained itself over the period.

TABLE 3.1 : **Size distribution of holdings—rural sector**

Size Groups (in Hectares)	1954-55		1961-62		1971-72	
	No. of Holdings	Operated Area	No. of Holdings	Operated Area	No. of Holdings	Operated Area
1. Less than 1.0	50.7	5.6	39.1	6.9	45.3	9.3
2. 1.0—2.0	16.9	10.0	22.6	12.3	22.7	14.9
3. 2.0—4.0	15.9	18.6	19.8	20.7	17.8	22.6
4. 4.0—10.0	11.6	29.2	14.0	31.2	11.2	30.4
5. 10.0 and above	4.7	36.6	4.5	28.9	3.1	22.8
Total	100.0	100.0	100.0	100.0	100.0	100.0

Source : Laxminarayan, H. and Tyagi, S.S. : "Some Aspects of Size Distribution of Agricultural Holding", *Economic and Political Weekly*, XI, 41, October 9, 1976, p. 1949.

* Figures are in % rounded to the first decimal point and are based on the 8th, 17th and 26th rounds of National Sample Survey.

The Draft Plan (1978-83) points out that up to the 1960s the land reform measures had no visible impact on the distribution of land holdings. As on 31st July, 1977, the estimated surplus area was 5.32 million acres, the area declared surplus was 4.04 million acres, the area actually distributed was only 1.29 million acres. Thus, the distributed area remains less than one-fourth of the estimated surplus. The result is that the general pattern of size and distribution of land holdings has not changed substantially over the years. Thus, one can conclude that land reform policies have not been able to make a dent in the rural power structure which continues to be based on inequalities of land holdings and lack of security to tenants and sharecroppers.

Community development and Panchayati Raj

Together with land reform, came the programme of Community Development initiated in the First Five Year Plan in 1952. In practical terms, the concept and organization of Community Development Programmes was influenced by several experiments already undertaken by the British Colonial Administration. However, the roots of the idea of community development can be traced to the work of Mahatma Gandhi who was associated with various programmes of upliftment of rural masses and who laid emphasis on the fundamental structure and integrity of the Indian villages. His ideas of "constructive work" were expanded by various people who established "projects" to show the utility of the theme. Many of these Gandhian ideas about the worth of the masses, village industries, village decision-making etc., came to be incorporated in the present national community development programme.

Most of the scholars evaluating the Community Development Programme found the inadequacy of the governmental machinery itself as the major obstacle to its success, (Taylor, 1956). It was by and large accepted that the civil servants were intelligent, hard-working and conscientious people but they lacked the essential spirit underlying the programme. Too much reliance on Government had also not helped in generating people's participation, (Dube, 1956).

Programme Evaluation Reports of the Planning Commission also pointed out the inability of the programme to help the economically handicapped classes. They showed that those who are benefiting from the development programmes and improving their economic and social conditions usually belonged to those sections of the villagers who were already some what better off than others in the village (Evaluation Report, 1957). The result of all this criticism was the acceptance of the scheme of democratic decentralisation or Panchayati Raj, as recommended by the Mehta Committee.

Panchayati Raj with its elective institutions was aimed at providing popular support to rural develpment programmes of the government. Balwant Rai Mehta Committee suggested a three-tier structure in which the elected bodies were inter-locked in membership allowing only one entry point at the base level.

The institutional framework of Panchayati Raj followed the pattern of Community Development Scheme. The roles of the District Collector, Block Develoment Officer, Extension Officers, and the Village Level Workers were unaltered. In addition, they were now expected to work with the non-officials too. Thus, the vertical hierarchical pattern of bureaucracy was maintained while a new set of people and institutions were introduced to control and guide them.

One of the consequences of this inter-locking web of relationships between the officials of the community development institutions and the non-officials of the Panchayati Raj Institutions has been the emergence of friction between the two and their lack of ablity to work as a homogenous team. (Bhambhri, 1967; Chaturvedi, 1964; Narain; 1966). The elected heads of these bodies have attempted to influence the officials to maximize the benefits for their own groups or their own areas. In these efforts, they have been helped by their linkages with state level leadership and their membership of the ruling party. The administrators, on their part, have resented the injection of this kind of leadership and have perceived them as threats to their own power and influence, (Mathur, 1972). This mutual hostility has often resulted in lack of concern about achievement of targets and inability to relate programmes to the good of all the people.

The overlapping relationships of Community Development and Panchayati Raj has also led to confusion in determining administrative boundaries among the various sub-systems within the district. For one thing, the officials have been placed under dual control. This control over the technical staff of the elected bodies is exercised by the Block Development Officer, President of

Panchayat Samiti, and the district technical officers of the Government Departments, (Dubey, 1972).

The major contribution of Panchayati Raj has been in the socio-political sphere. The fact that elections with the possibility of gaining access to decision-making bodies were brought to the village where traditional leaders uninterruptedly dominated was enough to bring in radical changes in both political processes as well as structure. Power has shifted gradually to the groups who were numerically more powerful or has led to formation of new groups that could maintain power. Thus, politics of elections has introduced the game of numbers. It is no longer possible for a numerically small group to hold power unless it generated support to become the majority. However, this has not really meant that those already holding power have been displaced; but what has resulted is at least the recognition by them that lower caste groups are important if they want to stay in power. On the other hand, the depressed groups have increasingly realized the significance of their vote which has led them to question their role on the periphery of the political system.

However, one should not make very much out of this kind of political revolution. For, due to their poverty and depressed social status, the under-privileged have really been kept out of the advantages of democracy and development. Recognition of their bargaining position has not resulted in the crystallization of favourable policies towards them or in getting a better deal from the State.

The picture of Panchayati Raj that emerged in course of its working was the very negation of the ideal that was held by the policy-makers. The image in which these institutions were established was a reflection of idealism and the Gandhian concept of Ram Rajya based on self-sufficient, self-dependent village communities. The whole argument of rural development and growth seemed to be raised on to a sublime level of morality and ethics. It was assumed that the villages would be free of any political or social struggles and problems would be solved through general agreement and consensus. Rural life was idealized and entry of politics into this area was sought to be restricted on a moral basis. However, the social-political reality predicted otherwise.

Villages were not confiict-free but were ridden with group rivalry and factionalism. There were family disputes, caste disputes, etc. Caste Panchayats that existed before were heavily influenced by the dominant group and resolved conflicts keeping their interests in mind. Introduction of Panchayati Raj brought these conflicts into open and allowed the groups to seek support even from out side the village to fight their point of view. The politicization obviously transformed local or caste issues into political issues which no longer could be solved through consensus.

Insistence on consensus in this context really meant maintenance of the power of the dominant group. Thus, despite the idealism and the demands of keeping politics out of Panchayati Raj, these institutions have helped in crystallizing political issues and polarize the village community, (Mathur, 1977).

In conclusion, it is important to point out that the professed goals of these institutions have begun to recede into background. The Panchayati Raj institutions have not been able to raise their own resources and have increasingly relied on government funds. An important consequence of this has been the increasing role of the government in implementing programmes. Rural leadership has seen its role in maximising the benefits of development for their area or their group and not in becoming enlightened base for the national policy makers talking in terms of social justice and increased agricultural production.

Cooperatives and village uplift

In an effort to develop village communities, one aspect of the effort was the initiation of Community Development and Panchayati Raj, the other was the emphasis on the village cooperatives.* Apart from seeing cooperatives as business propositions and doing what an individual can not do, the cooperatives were also seen as a method of mobilizing the people towards self-help

*Most of the ideas about the role of cooperatives can be traced to the way the whole movement developed during British India. As early as 1904, Cooperative Credit Societies Act, had been passed which broadly laid the foundation of the pattern of village cooperatives in India. There was great emphasis on credit societies and role of the state in promoting them. From then on, cooperatives became expression of government policy and not the urge of the people.

and village management. It was also assumed that cooperatives would be able to help the rural poor and others who are unable to take credit to improve their farm practices or who are kept out of development benefits because of the dominance of the traditional caste groups and land owners. To a great extent, therefore, cooperative credit programmes were initiated to protect the weaker sections from the exploitation of the money-lender and also in the procees help them through greater availability of credit to increase agricultural production.

However, the Cooperative Movement failed in general to fulfil the hopes placed in it. The weaker sections of the society whose need for their services is most urgent, have suffered most where these institutions have failed and did not stand to gain where they proved successful, (Khan and Tripathy, 1972).

In this failure, one responsibility has been on the kind of policy that has been evolved. Cooperative loans have generally been given on the basis of land though the Fourth Plan advocated that there should be increasing shift towards a policy such that loans could be given on the security of groups. Even if this policy could be implemented to give loans on the basis of producers of crops rather than owners of land the share-croppers and smaller farmers may not benefit very greatly because their produce is not very much and their surplus has already been extracted by the landlord. In any case, the landless labour are totally excluded from advantage, (Hunter, 1969).

In another contradiction in policy, cooperatives have become an administrative fabrication and not a symbol of cooperation. Because of official policy, administration has actively encouraged the formation of cooperative societies and the number established is the sign of the target achieved. The whole dynamics of co-operation is thereby lost; cooperatives are treated as another branch of bureaucracy and their secretaryship another job.

The All India Credit Review Committee (1969) further points out that the principle of open membership is not always effective for several cooperatives operate as closed shops for the benefit of a particular economic group or caste or faction. This again happened because of the bureaucratic approach. Government aid came out in a big way to support the cooperative efforts. Thus, a cooperative society emerged not because there was need for it but because some group had fulfilled the minimum requirements to demand funds from government. In such a situation. those persons who could manipulate rules and could foresee clearly the role of the cooperative for their own benefit took hold of existing cooperatives or established new ones.

The result was that the cooperatives fell prey to the powerful vested interests who were not essentially cooperators but politicians and ambitious of exercising influence in more than one sphere. At the state level, the state governments showed keen interest in maintaining hold on cooperatives but did not find any scope for initiative or any feeling of involvement in the working of the institutions with which they were associated. Concomitantly, with liberal financial support from government and wide network of societies spread throughout the State, ths apex institutions became strong sources of political power and prestige. Political linkages were quickly forged and the leadership looked upon cooperatives more as source of power than anything else, (Nicholson, 1973).

The new agricultural strategy and the green revolution

The rural development policy during the fifties, as we have seen was focussed on providing such institutional support that could modernize the village community and make it self-sufficient. It was assumed that the major difficulty with the rural people was their backwardness and lethargy and once their attitudes were changed than everything else would follow. Thus, inspired by an idealistic image of the village fostered by the Gandhians, institutions were planted in the rural scence in the hope of changing the people's minds and souls. Though everyone talked of increasing agricultural production but little direct and explicit effort was made in that direction. For, it was further assumed that all these expected changes will on their own lead to increased production.

A consequence of these assumptions was the policy of spreading the resources as widely throughout the country as possible. Targets were set in terms of setting up New Community Blocks and achievements were, therefore, measured in terms of the numerical increase of the Blocks every quarter. Because each block meant allocation of financial resources from the Centre and the States to that particular area demands for establishing more and more new blocks grew. Similar process worked for the coverage of the country by Panchayati Raj Institutions and cooperatives.

In this period of national enthusiasm for the work being done in the rural areas, it was gradually becoming obvious that agricultural production was lagging behind. It did not need much data to show that a crisis in agricultural production was approaching. Thus, the Third Plan was taking shape when it was not only being argued that agriculture needed much greater investment but also that "the rural sector could no longer be considered a welfare or social reform problem but rather an alternative to industry as a source of resources and growth" (Modernising Indian Agriculture Report, 1967).

This concern for agricultural production dominated the minds of the policy makers when the Ford Foundation presented its Report on "India's Food Crisis and Steps to Meet It" in 1959. The Government of India accepted in general the recommendations made in the 'Food Crisis' Report. The idea of Intensive Agricultural Development District as part of recommendations of the above report was accepted and the Government subsequently invited the Ford Foundation to assist it in planning and implementing this programme of increased agricultural production. The programme that was finally accepted with the advice of Ford Foundation was basically administrative in character, concerning itself with choosing certain districts only for intensive efforts towards greater agricultural production.

This approach of selecting few areas for concentrated attention received added strength when Mr. C. Subramaniam, the Food Minister announced his new agricultural strategy in 1965. The new agricultural strategy was based on the use of new seeds and application of technological advances in areas which were assured of irrigation. Mr. Subramaniam argued that the traditional varieties of seeds had little potentiality of higher yields and extensive use of high yielding varieties imported from Mexico and Taiwan must be made. It was also argued that the performance of the technology would be improved considerably if and only if applied to irrigated areas. Therefore, to maximize the benefits from technological advances, the Agricultural Ministry sought for certain concessions in the ideological goals that were focussed in bringing greater social justice in the rural sector. It was pointed out that the All India coverage of Community Development Programme tended to dilute scarce resources below the critical level necessary for increased agricultural production. Much better results could follow if these inputs were concentrated in the 22 per cent of the cultivable area where irrigation facilities were available. It was further argued that very high growth of output in this area would push up the overall rate of growth of agricultural production to a reasonably high level,

With the primary focus on increasing agricultural production, the new strategy departed from the ideological goals of the Indian agricultural policy. The whole idea of Intensive Agricultural District Programme and the High Yielding Variety Programme was to concentrate only on certain favoured areas in the country. This favour was extended to those areas in the country which boasted of irrigation facilities. This automatically restricted the choice to about 22 per cent of the cultivable area. The result of such concentration was diversion of resources from many other programmes to these efforts and one rapidly saw the withdrawal of funds from the Community Development and Panchayati Raj Schemes. In fact, many Chief Ministers began to talk of giving up these programmes totally. Most States, however, have reduced their outlay on the Community Development Programmes to levels sufficient only for staff maintenance. To a great extent the new strategy signalled a considerable dilution of the earlier approach of increasing agricultural production by giving social incentives to the farmer through far reaching institutional changes. The urge to look for a "progressive farmer" who would be motivated by resource concentration and price incentive was a tacit admmission by the leadership that they had failed to make the small producer an effective vector of economic progress through far reaching institutional reforms, (Francine-Frankel, 1969).

Thus, by and large, green revolution has been confined to limited areas with irrigation facilities and to those farmers who have the resources or access to institutional finance. Most small farmers, who lack resources of their own or have no access to institutional finance are outside the mainstream of green revolution, (Lipton, 1972).

Special schemes

Considerations of equality and justice and greater production have prompted the Government to initiate several new programmes specifically for the weaker sections of the rural society during the Fourth Five Year Plan. In pursuance of this objective, the Plan provided for two sets of projects. One was for small but potentially viable farmers and the other for marginal farmers and agricutural labourers. The small farmers were taken to be those who, with the help of irrigation, provision of credit, supplies and marketing and application of new technology, could be made viable. A limit of one to three hectares of land-holding was generally adopted in identifying such farmers. Farmers with land-holdings below this limit were classified as marginal farmers.

Under the programme, two new agencies Small Farmers Development Agency (SFDA) and Marginal Farmers and Agricultural Labourers Agency (MFALA) were initiated around 1970. These Agencies are incorporated in the Societies Registration Act and worked at the district level. Each agency has a small membership consisting mainly of the representations of the institutional agencies and concerned district officers. The District Collector works as the Chairman of the agency. While the Fourth Plan visualised 45 SFDA and 40 MFALA projects, the number of projects actually started were 46 and 41 respectively.* These 87 projects are spread over 97 districts, in part or in full. (Steering Group on Fifth Plan Report, 1973).

The SFDA's main tasks were to identify constraints facing the small farmers, provide small irrigation schemes, introduce new seeds, inputs and livestock farming and encourage a flow of credit at a subsidized rate. The MFAL Agencies were designed to help the marginal farmers as well as landless labourers through rural works programmes, homestead land and assistance for house construction. The marginal farmers were to be encouraged in dairy farming, small livestock rearing, fisheries etc. A recent evaluation (PEO, 1979) has found that only about 2.3 million families have been affected by the two schemes. Only a quarter of households had been identified and the number of beneficiaries has been even much less. Thus, the coverage of the programme is too

*These two schemes were merged in the Fifth Plan.

meager to make any significant dent on rural poverty. Problems arise from the nature of government working, credit institutions and the structure of agrarian society.

In addition to SFDA/MFALA, numerous specialized agencies have been established in the districts. All of them are set up to serve the needs of a specific area or a target group. In 1978, out of 392 districts in the country, only 62 districts have not been covered by one programme or the other. The distribution is as follows :

TABLE 3.2 : **Distribution of special schemes among districts**

Programme	No. of Districts covered
SFDA/MFAL	198
Drought Prone Areas Programme	74
Integrated Rural Development	20
Command Area Development	124
Tribal Area Development	64
Hill Area Development	3

Source : J. Sharma and Mehra, 1978, 3

What is interesting is that there is more than one programme operating in all the above districts.

Apart from these Special Schemes, the government intervenes through Employment Guarantee Schemes or Food for Work Programmes also. Thus, within the broad perspective of economic planning in the country, policies towards rural development have diverse objectives. There has been a perceptible shift in emphasis on policies since the days of enthusiasm for community development approach. In the last decade the concerns of new agricultural strategy have shaped most programmes and institutions for development of the rural areas. But the contradictions inherent in this approach of choosing selected areas and groups of farmers for increasing agricultural production and that of improving the production base of the deprived sections of rural population were recognised by the Government in the Fourth Plan itself (GOI : 149). It is this recognition that led to the strategy of formulating special scheme sand establishing specific institutions to reach out to such groups of people and such areas not effected by the new agricultural strategy.

The complex objectives of rural development policy and multifarious institutions chosen to achieve them placed a new burden on administration particularly when it was being relied on as the major vehicle of change. This burden at the field level where programmes were being implemented was particularly great with proliferation of programmes ; implementing agencies multiplied. The district administration presented several administrative hierarchies at work. The traditional District Collector heading the revenue and law and order set up continued but found himself formal head of all other development agencies at work in the district. He became the head of the developmental bureaucracy that came with the Community Development and Panchayati Programmes. He became chairman of the agencies that were introduced to implement special schemes. In addition he continued to coordinate the activities of all sectoral departments who had field offices in the district.

The major problem that emerged at the district level was therefore of coordination. It emerged because when policies or programmes were initiated enough attention was not paid to plan their implementation. Disparate agencies formulated discrete schemes which were then implemented by their own sectoral hierarchies. These had to be integrated to serve the ends of the district as a spatial unit. But departmental loyalties created great impediments in such integration.

Also, as the work, the District Collector was supposed to do was in continuous flux, he chose to establish his own priorities that he thought demanded his attention. This choice depended on several factors including those that related to his own personal predilections and those that reflected the socio political situation. What all this meant was that the district administration lived with multiplicity of programmes, goals, administrative institutions and hoped that some how they will be integrated and coordinated under the broad umbrella of the District Collector.

4

Green Revolution in Karnal

A MAJOR element of the new strategy that brought in Green Revolution is the package approach. New high yielding varieties, fertilizers, pesticides, appropriate water supply for irrigation and mechanical equipment together form part of a package. The core of the package is the high yielding variety seed which demands proper dose of fertilizer, water, pesticides in order to produce to its full potential. If these doses are not available at the appropriate time, the miracle seed cannot produce as expected. In addition, it also demands a particular quantum of the dosage. Thus, for the package to work fully, not only have the inputs to be provided but have to be utilized according to appropriate quantum and at appropriate time. Consequently, the package places heavy demands on the organizational system at the field level for making appropriate quantities of input at appropriate time available to the cultivator.

Another element of this new strategy, as we have noted earlier, is that it is selective in its approach. The package is distributed in those areas which are already well endowed with certain infrastructural facilities particularly that of irrigation and within those areas those farmers are favoured who are best able to utilize it. These farmers are by and large known as 'the progressive farmers'.

District Karnal in Haryana, a state in India located north-west of Delhi, has been one of those favoured areas in the country which form part of what is known as the Green Revolution belt. In the past decade or so, Karnal has registered a phenomenal

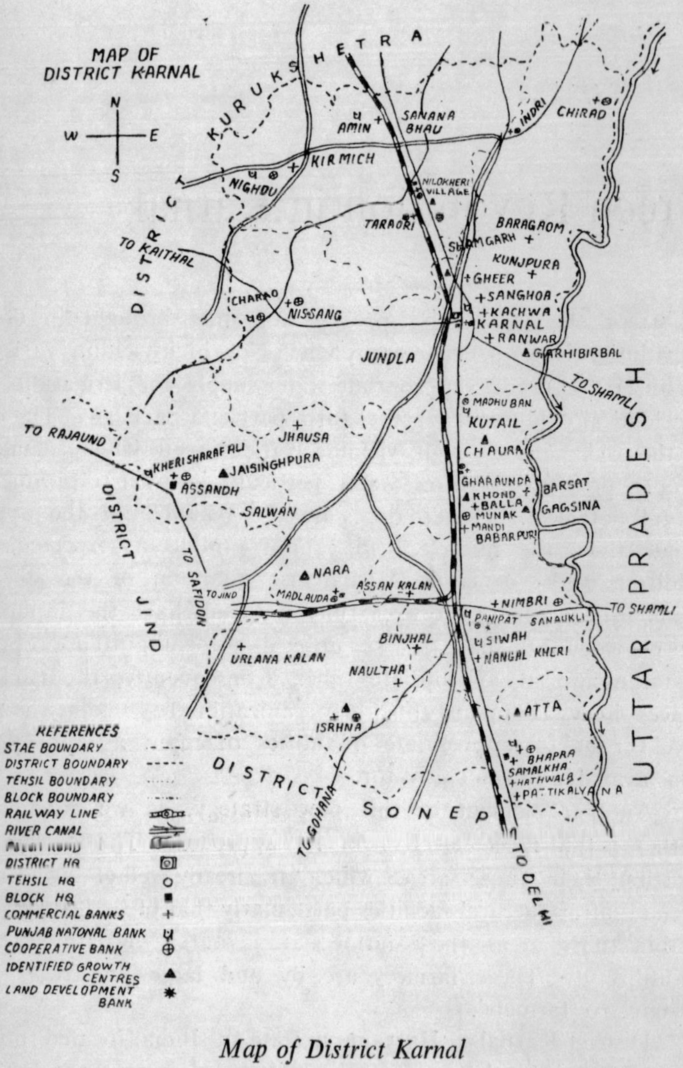

Map of District Karnal

rise in agricultural production and has made remarkable progress in paddy production particularly when rice is not a staple food of the people in the district. In this chapter, we intend to describe the pattern of growth that has taken place in this district and how this reflects the overall policies adopted for introducing the new strategy.

The total population of Karnal, according to 1971 Census, is around .99 million and the geographical area is about 3,617 sq. kms. The rural population is 79.44 per cent of the total and about 17.55 per cent of the people belong to the scheduled castes. What is important to note is that Karnal is well endowed with infrastructural facilities. It falls on the National highway which splits the district into two and is easily accessible from the national capital Delhi or the state capital Chandigarh. Pucca roads connect 84 per cent of its villages and 50 per cent of them are linked through government bus routes. All the villages are electrified. Further, out of eight community blocks in the district, five have their headquarters on the National highway itself. The road network provides easy access to markets, and communication with the outside world and places Karnal in a privileged position.

Occupational pattern

As Table 4.1 reveals, Karnal is primarily an agricultural district with about 42 per cent of the total workers being cultivators and another 22 per cent being agricultural labourers. The state of Haryana also reflects a similar feature but in a different way. The proportions of cultivators and agricultural labourer differ for the state and the district. In comparison to cultivators, the proportion of agricultural labourers is greater in Karnal than in the whole state of Haryana. All other proportions of occupational strata do not differ much.

There is another significant factor that needs to be pointed out here. The proportion of cultivators has declined during the period between 1961-1971 while that of agricultural labourers has gone up. As can be seen from Table 4.2 this has been true of the state as well as of the district.

TABLE 4.1 : **Working population of Karnal and Haryana : Census 1971**

Workers		Karnal %	Haryana %
Cultivators	—	41.65	49.08
Agricultural Labourers	—	22.01	16.21
Mining Quarrying	—	.93	1.54
Manufacturing, processing	—	10.47	9.96
Construction	—	2.38	1.84
Trade and Commerce	2	7.99	6.72
Transport, Storage	—	1.96	2.43
Other Services	3	12.61	12.21
Total		2,65,775	26,53,993

TABLE 4.2 : **Cultivators and agricultural labourers in Haryana and Karnal - 1961 and 1971 Census**

	Haryana		Karnal	
	1961	1971	1961	1971
Cultivators	63.87	49.15	57.67	41.65
Agricultural Labourers	6.90	16.21	9.72	22.01

* Proportion of cultivators and agricultural labourers is derived from total workers in the state and the district.

Source : *Statistical Abstract of Haryana*, 1977-78; 1966 (Economic and Statistical Organisation, Planning Department, Government of Haryana, Chandigarh)

Land use

Of the total geographical area, about 82.78 per cent of land was under cultivation in 1976-77, as compared to 77.81 per cent in 1966-67 in Haryana. For Karnal, the comparable figures were 80.37 per cent and 71.09 per cent. Thus, land under cultivation is going up. With this expansion in acreage, the intensity in cultivation is also rising. In Karnal, the area sown more than once, has gone up from 25.04 per cent to 51.83 per cent during the same period. Fallow land has been decreasing quite sharply thus effectively demonstrating that pressure on land has been rising.

The principal food grain crops of Haryana are wheat, rice, maize and bajra. As one can see from Table 4.3 it is not only

TABLE 4.3 : **Area under major crops and under high yielding varieties Haryana and Karnal 1976-77 and 1966-67**

Major Crops	Area as a Proportion to total Area under Foodgrain Crops in Percentage				Area under High Yielding Variety Seeds as a Proportion to total area under Foodcrops in Percentage			
	Haryana		Karnal		Haryana		Karnal	
	1976-77	1966-67	1976-77	1966-67	1976-77	1966-77	1976-77	1966-67
Rice	7.9	5.5	26.9	19.6	4.6	0.1	14.0	0.1
Wheat	32.5	21.3	52.9	34.4	28.9	2.5	51.0	6.8
Maize	3.0	3.0	7.6	8.1	0.5	0.1	0.8	0.1
Bajra	22.8	22.4	3.4	6.8	6.0	0.9	0.8	0.4
Area under Foodgrain Crops (in 000 ha.)	4148	3958	357	703	4148	3958	357	703

the area that has increased but in consonance with the strategy of the Green Revolution, the area under High Yielding Variety of seeds of these crops has increased substantially. The tremendous difference in proportions between the state and the district show Karnal as concentrating the impact of Green Revolution in this regard. Under rice alone, Karnal comes second after the neighbouring district of Kurukshetra. It is also noticeable that the major impact of the new strategy has really been more on wheat than on rice. The area under high yielding varieties of wheat was about 51.0 per cent while that of rice was 14.0 per cent of the total area under foodgrain crops in 1976-77. What is remarkable is while nearly all the area under wheat is under high yielding varieties, there is quite a difference between area under rice and under high yielding varieties of paddy. Only 52.04 per cent of area under rice is under high yielding varieties.

Pattern of land holdings

Before turning our attention to the characteristics of agricultural production, it is important to look at the pattern of land holdings in Karnal. The average size of operational holdings is about 4 hectares in Haryana and around 5.73 in Karnal. As can be seen from Table 4.4 there are more than 21 per cent of holdings that are above 5 ha. in Karnal but there are 24 per cent such holdings in Haryana. The proportion of operational holdings below two hectares is more or less the same in the district as well as in the state and is around 46 per cent.

TABLE 4.4 : **Operational holdings in Haryana and Karnal 1961 and 1971**

Class	Haryana %		Karnal %	
(in ha.)	1971	1961	1971	1961
Less than 1	27.37	6.25	27.97	6.26
1.0–2.0	18.90	9.90	19.66	9.12
2.0–4.0	29.36	41.03	30.94	46.95
4.0 and above	24.37	42.81	21.41	39.68
	(100)	(100)	(100)	(100)

Source : Statistical Abstract of Haryana, 1977-78 : 1971-72.

This means that around 30 per cent of the holdings are in the range of 2-4 hectares. The data of 1971 presents a striking comparison to that of 1961 when the proportion of farmers holding less than 2 ha. of land was considerably less both in the state as well as in the district. This means that during the decade 1961-71, two simultaneous processes took place. The proportion of agricultural labourers increased together with the proportion of those who owned less than 2 ha. of land. In addition, the proportion of holdings above 4 ha. has decreased considerably during the period 1961-71. There appears to be a process of downward push and probably many of the small landholders owning up to 2 ha. have been rendered landless.

Irrigation pattern

Availability of water supply forms the core support in the adoption of high yielding varieties of seed. During the period 1966-67 to 1976-77, the area irrigated in Haryana increased from 29.6 per cent to 43.3 per cent. During the same period, in Karnal, the area irrigated increased from 45.33 to 77.90 per cent. But the prominent characteristic in Karnal is that the area irrigated by canals has declined. Tubewells dominate the scene today and account for about 79 per cent of the total irrigated area. This has happened for two reasons. First of all ground water resources are available in Karnal because the canal system has helped to raise the water table which can be exploited economically. But secondly and for more important reasons, cultivators find that they have a greater degree of control of water supply to the fields if they have tubewells to irrigate their fields. This control aspect of tubewell irrigation is particularly important for paddy growers who demand water at a particular time in specific quantities. However, the major constraint in effective use of tubewells is the availability of power. Tubewells in Karnal are operated both on diesel as well as electricity and during 1976-77, there were 8,969 diesel and 29,351 electric pumping sets in the district. During the period under study, both sources of power, diesel as well as electricity, were under short supply.

The availability of credit facilities to individual farmers for installing tubewells was a great incentive. The minor irrigation scheme sanctioned by Agriculture Refinance Development

Corporation in 1967 provided for the installation of 3,600 shallow tubewells in the eight blocks of Karnal, Gurgaon and Ambala districts of Haryana. The total financial outlay was that of Rs. 2.70 million. The target was achieved in 1972. Seventy per cent of the tubewells were in Karnal district.

A major feature of tubewell irrigation is that it is largely privately owned and due to the amount of financial investment required, only large farmers who have the capacity to raise loans of this amount can finance it. In an evaluation study of minor irrigation scheme sanctioned in Karnal by the Agricultural Refinance and Development Corporation (Reserve Bank of India, 1977), it was found that the largest single group of borrowers who availed of the loan facility of Rs. 7,500 (estimated cost of tubewell etc. during 1967-68 to 1971-72) was having cultivated holdings ranging between 4 to 8 hectares (p. 35). In recent years the cost has increased considerably and would be higher in areas where the level of underground water is lower. The study then goes on to add, "though the state government was to ensure that the small farmers were not excluded from the scheme of development, only a small number of such farmers appeared to have participated in the scheme" (Reserve Bank of India, 1977 : x). An important reason for the inability of the small farmers to participate in the scheme, apart from what the study mentions as lack of effort by the bank concerned, was due to the clause that the "amount of loan could be sanctioned against 50 per cent of the value of mortgaged land" (p. 12) and most cultivators in the sample studied appear to have given 2-3 hectares of land in security. Small farmers understandably were unable to avail of the facility provided in the scheme and their poor response could not only be due to promotional efforts of the state.

Credit facilities

A major characteristic of the new agricultural strategy is that it makes the farmers increasingly dependent on credit for supply of inputs. These inputs are expensive and the cultivator cannot rely on last years savings alone to buy them. Besides, multiple cropping demands a constant flow of investment without waiting for returns on one crop to come in. Such a flow cannot be maintained without credit. Traditionally, the money-lender supplied credit but because of exorbitant rates and exploitative character of lending he was found inadequate. Credit through cooperatives was introduced primarily to replace the money-lenders and release the farmers from their clutches. However, cooperatives have continued to exist side by side with money-lenders. Such an arrangement was not found suitable to fulfil the demands of the new agricultural strategy.

The government has sought to introduce a multiagency approach while maintaining the cooperative credit system. The cooperatives comprise two types of institutions. For short-term credit not exceeding five years given for seasonal crop inputs and for marketing there is a District Cooperative Credit Bank with primary societies at the village level. At present there are 15 branches of the District Cooperative Bank spread all over the district. There are 186 primary societies in the district and are called mini banks. The long-term credit is provided through Primary Land Development Banks which have two offices in the district.

In addition, after the nationalization of banks, commercial banks have been asked to advance loans for agricultural purposes. Under lead bank schemes, a commercial bank is given the responsibility of developing credit schemes for one particular district. Punjab National Bank has prepared a District Credit Plan for Karnal. Guidelines from the Reserve Bank of India are issued for giving loans for such purposes as crop production, tubewells, tractors etc. Other types of credit schemes are those which have emerged from the participation of Banks with World Bank projects.

Thus, loans are advanced to farmers through cooperative credit societies known as mini banks, primary land development banks and commercial banks. The cooperative loaning is restricted to short-term loaning with a little bit of medium-term loaning.

With the introduction of commercial banks in the agricultural credit structure, the Reserve Bank of India sought to guard against the out flow of deposits mobilized in the rural areas. Therefore, it required that the commercial banks maintain credit deposit ratio at 60 per cent at corporate level. Figure 4.1 shows the credit deposit ratios during 1972-77 for rural, semi-urban and urban areas of district Karnal. It can be seen that rural credit/deposit ratio has been rising during 1973-75 and then declining after that. Probably the rise was due to the World Bank credit scheme administered through Banks for farm mechanization during that period. Subsequently, advances on this account slowed down and hence the credit/deposit ratio also started decreasing. In 1976-77, according to the District Credit Plan, it was 52.25 per cent.

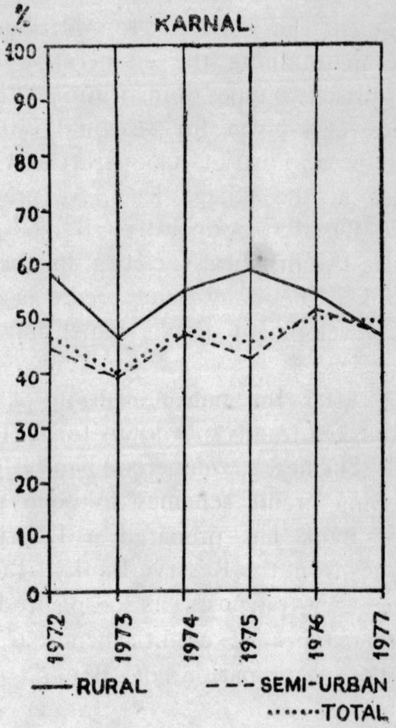

FIG. 4.1 : *Credit deposit ratios 1972-77*

Progress in rice production

With all these facilities being provided, the contribution of Karnal for increased agricultural production in the state is spectacular. A recent study (Bhalla & Alagh, 1978) sponsored by the Planning Commission, Government of India has shown that between 1962-65 and 1970-73, Haryana was among those states in the country which have registered the maximum increase in output per hectare. Karnal is among the 28 districts in the country which have recorded growth rate of above 4.5 per cent with the yield level rising between Rs. 700-1300 per ha. during the same period.

Among foodgrains wheat has done better than rice. The last report of the Agricultural Prices Commission (1978-79) mentions that the All India production of rice increased by 7 million tonnes or by 17 per cent from an average of 40.8 million tonnes in the year ending 1977-78. Of the additional production, as much as 42 per cent was accounted for by Punjab, Haryana and Western Uttar Pradesh. As a result the share of these three states in total production of rice in the country went up from 10.6 to 15.4 per cent. Production of rice went up by 120 per cent in Haryana.

The all India yield of rice went up from 1019 kgs. per hectare to 1215 kgs. during the period between triennium ending 1970-71 and 1977-78. The rice increase in yield in Haryana was 60 per cent and it achieved a level of 2392 kgs. per hectare. Another significant fact about Haryana's rice production is that in 1977-78 as much as 91 per cent of its produce was sold to government agencies. Apart from the fact that rice went to the market because of price, it must also be recognised that Haryana is not a rice eating area and has only recently in the past decade become a rice producing area. Therefore, rice is a commercial crop in Haryana produced for profit and not for consumption.

Karnal district has been in the forefront in contributing to this spectacular achievement in Haryana. Over the ten years period 1966-67 to 1976-77, the average yield of rice per hectare increased by around 67 per cent from 1214 kgs. to 2035 kgs. The production of rice went up from 2,23,000 tonnes in 1966-67 to 9,64,000 tonnes in 1977-78, an increase of more than 4 times.

The area under rice production went up from 5.75 per cent of the net area sown in Haryana in 1965-66 to 9.04 per cent in 1976-77. In Karnal, the proportionate increase was much larger from 20.51 per cent to 32.04 per cent. No doubt the growth rate in average yield per hectare has been much greater. What is remarkable, though, that area under maize and bajra has remained more or less static, (see Table 4.5) in spite of the fact that the yield level of bajra went up from 395 kgs./ha. to 1003 kgs./ha. Maize, however, did not register such an increase though the level went up from 874 kgs./ha. to 1067 kgs./ha.

TABLE 4.5 : **Area under principal crops as a proportion to net area sown, Karnal 1965-66 and 1976-77**

(*in percentage*)

Crops	1965-66	1976-77
Rice	20.1	32.1
Wheat	34.4	63.2
Maize	7.6	9.0
Bajra	5.5	4.0

Bases : Statistical Abstract of Haryana, 1966 and 1977-78.

Conclusions

There is clear evidence from the above description that Karnal has made spectacular progress in agricultural production in the last decade or so. The Karnal farmers have demonstrated their entrepreneurship by expanding the area under high yielding varieties and increasing the consumption of chemical fertilizers. Mechanization of farm activities has also kept pace with increasing use of tractors, diesel engines, electrict pump-sets and other agricultural machinery, irrigation facilities have also considerably multiplied and so also credit support.

In this section, we will attempt to delineate certain basic characteristics of the pattern of development that has taken place.

The differential impact of new agricultural strategy on wheat and rice signifies some special features of the kind of development that has taken place. Wheat and rice are grown in two different seasons, those of rabi and kharif respectively. Kharif crops are known to be susceptible to environmental conditions particularly rainfall. Even when irrigation is assured for nursery and transplanting activities, the photosensitive nature of paddy makes it vulnerable to cloudy days, humidity etc. Consequently, it can be seen from Fig. 4.2 that when rainfall touched a low ebb in the years 1969-70 and 1974-75 there was much greater variation in rice production than in wheat. Consumption of fertilizer also followed a similar trend. Such fluctuation was also marked in the relationship between harvest prices and rainfall (see Fig. 4.3).

What these trends show is that paddy farmer is more affected by the vagaries of nature and the fluctuations of the market, a further implication of these trends is that a farmer assured of full control over his irrigation facilities will be better able to minimize the uncertainties involved in rice production. Therefore, bigger farmers who have tubewell irrigation will be more disposed to growing paddy. This relationship comes out clearly in our own data presented in Chapter 6.

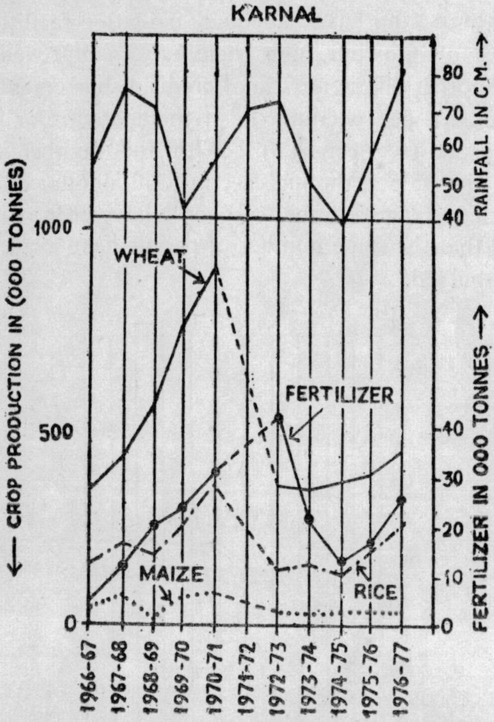

FIG. 4.2 : *Rain fall, crop production and fertilizer consumption in Karnal 1966-67—1976-77*

Credit has also flowed unevenly to different blocks in the district. The two sample blocks of Nilokheri and Samalkha, to be described in greater detail later, underline the linkage between flow of credit and irrigation facilities. Samalkha lacks financial facilities and there is no commercial bank in the block apart from Samalkha proper. The farmers have to depend on money-lenders and cooperative societies for loans. Nilokheri farmers enjoy far greater choice and availability of credit. One of the major differences between the two blocks is that the area irrigated by tubewells is 83.90 ha. in Nilokheri but only 53.93 ha. in Samalkha. The number of tractors and pump sets has followed the extent of irrigation.

In addition, the District Credit Plan (Punjab National Bank, 1977) itself mentions that the crop loan is generally availed of by those cultivators who have "adequate irrigation facilities and have an aptitude for growing high yielding or improved variety of crops". Another characteristic of credit disbursement is that the share of credit per account has remained more or less constant over the years (see Fig. 4.3). With the number of accounts remaining the same, the increase in total amount disbursed does not mean that greater number of people have been reached. What it means is that the same number of people have availed of greater amount disbursed.

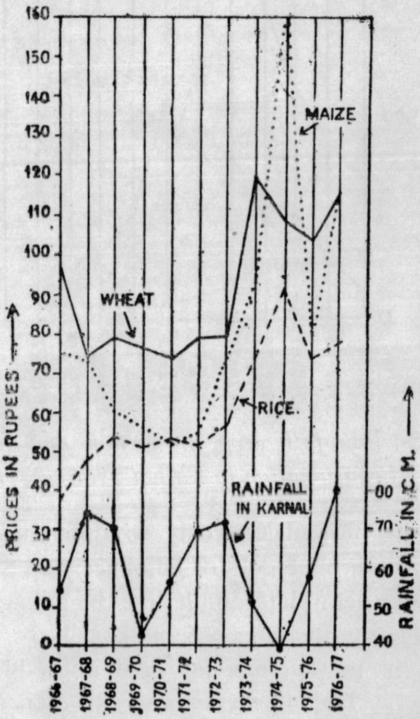

FIG. 4.3 : *Harvest prices and rainfall in Haryana, 1966-67—1976-77*

Another characteristic that seems to tie up with the above analysis is that tubewells are the dominant source of irrigation in Karnal. This has several implications. Tubewells provide greater confidence to the cultivator because he can use the water when he wants it—both in terms of timing and quantity. In contrast, canal irrigation provides less assurance and demands greater discipline on the part of the farmers. Channels have to be built and maintained and a time sharing system has to be devised. But as tubewells demand investment, there is a greater probability that they are owned by farmers who own larger parcels of land. The result is that only a certain group of farmers are able to reduce their risk and uncertainty. The situation is aggravated in case of rice production as already discussed.

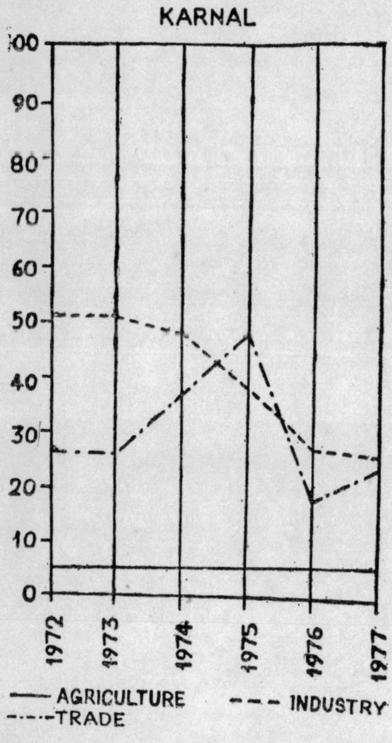

FIG. 4.4 : *Credit per account, 1972-77*

The overall picture that emerges is that of prosperity spread unevenly. Technological advancements have undoubtedly taken place but the gains of these advancements have not been distributed equitably—in terms of farmers or areas. There are also indications that some groups of farmers may have become poorer in the process. A decline in the number of cultivators has occurred with an increase in the number of landless agricultural labourers. There is a push from the top taking place and the people below appear to be worst affected.

5

Administrative Organization for Increased Agricultural Production : Headquarters and the Field

ONCE the new agricultural strategy was broadly accepted as a means of increasing agricultural production, a vast expansion of administrative organizations took place. Either new agencies were established or the existing ones were given more functions to perform. This process took place both at the State as well as district level.

At the state level, the administrative framework is very similar to what exists at the Central level. There is a Secretariat and also Directorates that are responsible for the implementation of specific policies. These directorates are spread from their state headquarters to the districts. In most cases they are staffed by professionals who are known as subject matter specialists. Each of these specialists are part of their own departmental hierarchy and governed by their own rules and regulations, career opportunities etc. In general, the highest rank that such subject matter specialists can attain is to become Directors of their departments like Director of Agriculture. Rarely do they rise to senior positions in general administrative hierarchy for such positions are usually held by members of the Indian Administrative Service—the Service from which the District Collector is always drawn.

At the district level, the implementation of these development programmes is carried under the broad coordination and guidance of the District Collector. One of the characteristics of the Indian development administration has been that while the field agencies of the technical Directorates at the district level have proliferated, the basic structure of district administration with the man from the Indian Administrative Service as the District Collector has remained unchanged. This continuation has also meant that he has remained in charge of the district and other functionaries or agencies recognize his pre-eminent position in the district. Consequently, the formal role of implementing agricultural development programmes falls on the Collector and the field departments through their technical and professional competence provide him support and assistance.

Organization of the directorate of agriculture in Haryana

The Department of Agriculture is headed by a Director of Agriculture in Haryana. He is assisted by an Additional Director of Agriculture (ADA), six Joint Directors of Agriculture (JDA) and a Deputy Secretary to the Groundwater Board. The Administrative Officer who is in-charge of establishment and accounts, also reports directly to the Director.

The ADA, JDAs and DS have been given specific responsibilities and are in-charge of activities grouped under one or more branches (see the organization Chart attached). Efforts have been made to avoid as much overlapping as possible. The following activities have been grouped under nine branches headed by Class I Officers like ADA, JDA or Dupty Secretary who report directly to the DA.

(1) Extension
- —General Extension
- —F.T.C. and training institutes
- —Field demonstrators, minikit trials, training campaigns and coordination with Haryana Agriculture University.
- —Information and publicity

(2) Marketing
- —Marketing
- —Warehousing and storage
- —Coordination with Agricultural Marketing Board

(3) Crop Planning
- —All food crops except pulses
- —Pulses

	—All cash crops (sugarcane, cotton, oilseeds)
	—Horticulture
(4) Plant Protection	—Plant protection
(5) Input Planning	—Fertilizers and manures
	—Seed and seed farms
	—Credit
	—Coordination with HAFED, with crop Bank, with HSDC
	—Seed certification on agency
(6) Land Improvement	—Soil conservation
	—Soil survey
	—Special projects under soil conservation
	—Coordination with LRDC
(7) Ground Water Utilization	—Ground Water Cell
	—Minor Irrigation
	—Coordination with MITC, ARDC
(8) Agricultural Engineering	—Gobar gas (cow dung) plants
	—Agricultural engineering
	—Coordination with Agro-industries corporation
(9) Planning and Statistics	

The organizational chart on the next page shows how these activities are divided up as responsibilities for several officers of the department.

Field organization

At the district level, the Agricultural Department is headed by a Deputy Director. With him there is an Assistant Plant Protection Officer and a Horticulture Development Officer. In the IADP districts, there are also other Officers like Agricultural Information Officer and Seed Analyst. Statistical sections headed by supervisors exist in all districts. Karnal is not only an IADP district but also the area where the World Bank project of Training and Visit System is being implemented. Therefore, a supplementary set of subject matter specialists (5) has also been added to the existing establishment. There are also farmers

Organisation chart department of agriculture headquarters

Director of agriculture-cum-Cane Commissioner

A.D.A.	JDA(P)	JDA(HYP)	JDA(SC)	DDA(QC)	JDA(PP)	DS(GWC)	JDA(AE)	AO
P.O. D.D. (STAT)	D.D.(H&V) M.O. A.I.O.	D.D.A.(HQ)	D.D.A.(SC)	D.C.C. D.D.A.(CC)	L.C.&P.O. P.P.O. A.S.P.O. F.R.O.	H/G W.D.S./EXM S.H/EXM W.D.S.	A.E.	E.O. A.A.O.

RESPONSIBILITIES

*Planning & Statistics	*Information & Publicity *Marketing *Horticulture *Agricultural Marketing Board *Ware housing & Storage. *H.A.U. *I.C.A.R.	*General Ext. *All Food Crops Except pulses *F.T.C. & Training Institutes *Field Demonstrations, Minikit Trials, Training Compaigns and its Coordination with H.A.U. *H.S.E.B. *Canals	*Soil conservation *Soil Survey *Special-Projects *L.R.D.C.	*All Cash crops (Sugarcane, Ollseeds, Cotton) *Pulses	*Fertilizers & Manures *Seed & Seed-farms *Credit *Seed certificate on Agency *H.S.D.C. *Plant-Protection *HAFED *Cooperative Bank	*Groundwater Cell *A.R.D.C. Scheme *M.I.T.C. *M.I.	*Agril. Engineering *Gobar Gas Plants *Agro Industries Corporation	*Establishment & Accounts

training centres spread in the district and are headed by a Chief Training Officer located at the district headquarters. Thus, Karnal district appears to be in a privileged position as far as the complement of departmental staff is concerned. It has more "specialists" than any "ordinary" district in Haryana.

The idea of Intensive Agricultural District Programme (IADP) emerged out of a Ford Foundation Report on India's Food Crisis and steps to meet it in 1959 (see Modernizing Indian Agriculture : Report on the Intensive Agricultural District Programme 1960-68 Vol. I, Government of India, 1969). The selection of districts under this programme was done under the following criteria :

1. The district should as far as possible have assured water supply.
2. It should have a minimum of natural hazards i.e. there should not be problems requiring long-term attention; susceptibility to floods, drainage problems, acute soil conservation problem, etc.
3. It should as far as possible have well developed village institutions like cooperatives and Panchayats.
4. It should have maximum potentialities for increasing agricultural production within a comparatively short time.

In administrative terms, the programme was bureaucratic in character. More technical officers were added and the project in each district was headed by a Deputy Director Agriculture who worked in close collaboration with the District Collector. Another significant change was the reduction in the work-load of the Village Level Workers. Known as Assistant Development Officers in Karnal, they were now asked to look after much lesser number of villages. Thus at the district level, under this set up, Karnal has a Duputy Director Agriculture, with Sub-Division Agriculture Officer, Plant Protection Officer, Agricultural Information Officer, Seed Analyst, and a Statistical Officer. There is also a Farmers' Training Centre headed by a Chief Training Officer.

The field level organization coincided with Block Administration till the Training and Visit System was introduced under a

World Bank Scheme. Under this scheme, several changes in the extension system have been introduced. Some of the major assumptions on which this scheme is based are the following: (a) Usually, the extension agent has to report to several supervisory levels and his links with the agricultural department are rather tenuous. In India, he has been part of the organization of Community Development and therefore is a multi-purpose worker. Most of the time he has been kept busy sending up reports and statistics, (b) Most field level extension agents have an excessively large jurisdiction. This task is rendered even more difficult because of the spread of the families to be contacted and lack of facilities for adequate mobility, (c) Training of extension staff is inadequate. Great store is set by pre-entry training but this knowledge is theoretical and even this is quickly forgotten in the humdrum of daily administrative work. Furthermore, when extension agents are regarded as multi-purpose workers, their training is further diluted by non-agricultural subjects, (d) There is usually no effective link with research activities. Without a continuous flow of practical recommendations suited to farmer's needs, the extension agent has not much to offer to the farmer.

The T & V System seeks to remedy these inadequacies in the traditional extension system. For this purpose, several changes have been accepted by the Haryana Government in implementing this World Bank project. At the district level in Karnal five subject-matter specialists in the area of agriculture, plant protection, soil and fertilizer, training and information have been introduced. They work under the overall charge of the Duputy Director Agriculture. The field level extension service has been reorganized by taking it out of the Community Development administration and placing it directly under the charge of the Agricultural Department. The district has been divided into 16 agricultural circles and placed under the charge of Circle Agricultural Officers (CAOs). Under each CAO there are 8-9 Agricultural Development Officers who are in direct touch with farmer families. There are 131 ADOs working in the Karnal district. Each ADO has around 750 farmer families in his direct jurisdiction. Usually, he determines around 10 farmers in each village as his contact farmers.

Together with this set up, there is also a parallel administrative system of the Development Department which retains Block Administration. There are 8 Blocks in the district. These blocks coincide with the boundaries of Panchayat Samitis which are the participative local institutions. At the village level there are 483 Panchayats. The district level body known as Zilla Parishad were abolished by the Government in 1973-74 and work is taken care of by the District Collector himself.

Administratively, the district is divided into two sub-divisions viz. Karnal and Panipat. Karnal sub-division consists of 9 agricultural circles with the rest in Panipat. However, Karnal sub-divison has 5 blocks while Panipat has three. Agricultural circles are administrative units for the agricultural department but in a few cases the circle and block boundaries coincide.

The discussion of government's organization for agricultural development remains incomplete without mentioning several corporations that the government has set up to work in this area. The Haryana State Cooperative Supply and Marketing Federation Ltd., (HAFED) is the apex body of 68 cooperative marketing societies in Haryana and its purpose is to provide agricultural inputs and facilities for marketing of farm produce and promote agriculture by setting up agriculture based processing units. In terms of inputs, it provides seeds, fertilizers and pesticides to the farmers through its outlets throughout the State and the district. The Haryana State Minor Irrigation Tubewells Corporation Ltd., does extensive work for exploration, development and exploitation of the underground water potential of the State. The Haryana Agricultural Marketing Board is concerned with marketing while Warehousing Corporation looks after the storage of agricultural produce. There are also many other corporations like Dairy Development Corporation, Rural Development Board, Agro-Industries Corporation etc., who play an important role as implementing agencies for the agricultural policy of the State.

The departmental organization of Haryana at the State level reflects the way most departments in Governments in India are organized. The organization is based on the assumption, (a) that the work of the department is divisible into a number of broad areas or subjects, (b) that consequently this work can be divided between a suitable number of senior officers, (c) that the impact of a decision taken in one area of work on another area of work can be worked out in consultation between senior officers, and (d) that this process of consultation is sufficient to provide the integrating force necessary to ensure that the various officers (and hence the work) do not operate in an uncoordinated fashion.

Such a model has important implications for organizational functioning. One of them is that tremendous significance is attached to the inter-personal relationships in ironing out difficulties and resolving problems. Power tends to be concentrated at the top of the hierarcy. Consequently, innumerable committees have to be formed and meetings held to have a continuous monitoring of the way work is proceeding. Every problem of coordination tends to be pushed up to these committees and to these meetings. The functionaries at the operational levels do not feel compelled to solve them with their colleagues or peers. It is only natural that a system like this will cause delays.

Another striking feature of the organization of agriculture department described above is that while the headquarters is totally absorbed in planning and setting targets, the field organization is only responsible for persuading the farmer to utilize the supplies and inputs so that the overall production targets are achieved. The department is not strictly involved in supply of inputs. For this the Government of Haryana has created several other organizations mentioned above.

These organizations have their functionaries located at district as well as sub-district level and are responsible for implementing their own sector targets like X amount of seeds, or X amount of fertilizer etc. that has to be distributed in a particular district. The departmental functionaries persuade the farmer to use a certain dose of fertilizer or a certain quantum of seeds. How the farmer gets them and from where is somebody else's responsibility.

Thus, the departmental functionaries at the field level like the ADOs have also to keep constant touch with implementing agencies who have the inputs. The result is that in a green revolution area, where the importance of adequate supply of inputs is widely accepted, the attention of the farmer turns towards those who can ensure it. The poorer farmer, in any case, is unable to gain credence of the extension agents because he does not have the capacity to accept inputs that are provided.

Another implication is in the area of performance. The Deputy Director of Agriculture in the district is handed down targets for distribution of fertilizers, seeds, pesticides etc. when he has no authority over the respective implementation agencies. What is being expected is that an officer in one hierarchy will be able to supervise the work of officers in another hierarchy. Such an expectation tends to ignore the tremendous role that authority and departmentalism play in Indian administrative life. The result is that the district functionaries can blame other agencies for not accepting the targets of inputs supplies and thus seek an *alibi* for their inability to fulfil targets. The channel to push up difficulties to the State headquarters is, of course, always open to them. They can then wait for resolution of their problems before proceeding further.

Most interestingly, a consequence of this way of functioning is that great importance is attached to meetings. While State headquarters hold meetings to plan, the district does the same to implement rather than plan—a district level coordination committee may have more than fifty officials in attendance. A flurry of activity goes on and anybody observing Indian administrators find them moving from one committee room to another and striking personal relationships so that work can proceed.

The picture that emerges of the operational machinery is that there is a multiplicity of organizations working at the district. Separate organizations have been formed to perform individual functions. By and large, each input is provided on its own organizational basis while the Department of Agriculture performs the task of extension. With this functional diversity being embedded in separate organizations, issues under the broad rubric of coordination emerge as the major problem area of implementation. This is compounded by parallel administrative systems running in the district. In this administrative proliferation, Panchayati Raj scheme of local level participation has been imposed. The result is that there is a blurring of lines of responsibility and areas of tension tend to rise.

6

Moneylenders, Institutional Credit and Green Revolution

THE fundamental basis for the acceptance of new technology has been the increased flow of credit to the farmers as we saw in the earlier chapter. Credit was essentially taken by those who not only had the capacity to utilize it but also the capacity to repay it. A farmer with a larger land holding becomes a natural choice for the credit agencies for becoming their customers. Secondly, institutional credit also sought to replace the moneylender who was the traditional source of credit in the rural areas.

At the village level, for short and medium term credit, the public institutions provide two forms of credit, crop finance loan and taccavi loan. It is the crop finance loan that has been used primarily to finance the high yielding variety programme. The taccavi loan is given by the administration usually at times of stress or physical emergency.

While the cooperatives look after short and medium term loans, the long term loans are disbursed by the land mortgage banks. With the nationalization of commercial banks in 1969, they also came in to provide credit for increased agricultural production. Such credit has increased over the years and has enabled the farmers to buy agricultural inputs from fertilizers to tractors, to buy land and to make improvements in buildings and to diversify their activities.

Even though the supply of credit has expanded considerably, it has been argued that the scale of crop finance is inadequate to meet the needs of the farmers, (Dasgupta, 1977; 113). Most farmers who have adopted high yielding varieties have done so on the basis of their own resources or from borrowings from private sources. But still they have tended to corner institutional lending. According to official estimates, the increase in debt of agricultural labour households was above all India average in Haryana and Punjab—the two most agriculturally prosperous states in the country. The average amount of debt of such a household was Rs. 1338 in 1974-75 in Haryana. Out of this while only Rs. 169 was owed to Cooperative Societies, the amount owed to moneylenders was Rs. 1025. (see All India Rural Labour Enquiry Report, 1974-75).

To capture the way the poorer sections of the rural community are trapped in business transactions with moneylenders we have looked into cases of three individuals specifically. Perhaps the three cases bring out the variances of credit flow more clearly.

1. Brahm Prakash

Profile

—Practically, a marginal farmer, has three brothers and two sisters, two brothers live and earn separately, while he and his younger brother live with their mother. Father expired long back.

—Major source of income a Cart-Bhainsa Buggi (Cart driven by a buffalo).

—Cultivates land of the village Bania (moneylender) on lease.

—Owns hardly 1/2 acre ancestral land.

Case

Brahm Prakash along with his brother had been plying the cart between the village and the market for several years. They carried goods of village shopkeeper and Bania and did other sundry jobs for him. In 1975, he took on lease one and a half acre of land from the same Bania for a year. The condition of lease was that an amount of Rs. 1100 would have to be paid by him after selling the paddy crop. In case of default Bania was to charge interest @ 3 per cent per month. Brahm Prakash borrowed Rs. 500 at the same interest from the Bania for buying the seeds and fertilizers also.

Due to shortage of canal water and lack of timely and adequate rains he could not have very good yield. He got 20 quintals of rice which was sold at the rate of Rs. 82 per quintal amounting to Rs. 1,640. The land preparation, seed and fertilizer etc., had already cost him Rs. 650.

He had also borrowed Rs. 300 in August 1975 for the purchase of buffalo since his earlier one had died. Thus, in December 1975, he owed the Bania Rs. 2,216 (including Rs. 1100 for land, Rs. 500 for seed and fertilizer, Rs. 300 for buffalow plus the interest). Bania after adjusting the proceeds from the paddy crop, that is Rs. 1,640, showed an outstanding balance of Rs. 576 against Brahm Prakash.

In 1976 he borrowed Rs. 1,900 with interest @ 3 per cent per month from the Bania for whom he carried goods by his cart. He had to borrow this amount for performing marriage of his younger sister. He then owed Rs. 1900 plus Rs. 576, i.e. Rs. 2,476 which he promised to repay after Rabi Crop along with the interest @ 3 per cent per month.

Brahm Prakash had not been able to use proper amount of fertilizer and good variety of seeds due to shortage of money with him and the heavy indebtedness. He had no sure source of irrigation. He could hardly get 18 quintals of wheat out of which he retained six quintals for domestic consumption and sold 12 quintals to Bania for an amount of Rs. 1200. By the end of the 1976 the original amount after adding interest had become Rs. 2,857. After adjusting Rs. 1,200, balance of Rs. 1,657 still remained to be paid.

Brahm Prakash had to further borrow for the *Gona* ceremony (formal sending of bride to her in-laws place after some period of marriage), an amount of Rs. 200 at the same rate of interest. He also took the lands of Bania on lease for Rs. 1100 for growing paddy crop. Another loan of Rs. 500 was taken for purchase of seeds and fertilizer. Unfortunately, soon after the transplantation of paddy in the month of July, younger brother fell ill. He had to borrow Rs. 700 for his medical care. He had thus to pay to Bania a total amount of Rs. 4,157 out of which Rs. 1100 for the land was without interest while the rest Rs. 3,057 was to be charged on an interest of 3 per cent per month. The paddy crop could not be attended to properly because of illness of his brother, and due to lack of water etc. yield was thus quite low. He was anxious about the mounting burden of loan and had no source of money to pay back. His debts were increasing. He got only 22 quintals of rice in Kharif season which were sold for Rs. 1,900 to the Bania who adjusted his Rs. 1,100 towards lease amount and gave Rs. 800 to Brahm Prakash for the domestic consumption. After compounding interest, by the end of 1976, his debt stood at Rs. 5,000.

He did not take any loan this time but in July 1977 the amount had grown to Rs. 6,000. The wheat yield he got in Rabi was worth Rs. 1000 only and this amount had to be paid to Bania.

Bania demanded the payment of remaining balance also. Brahm Prakash expressed his inability and requested for some more time. Brahm Prakash had hardly half an acre of land with him which he had inherited from his father. Bania suggested disposal of this land. He sold quite reluctantly this land to Bania and adjusted the loan amount.

This was not the end of his indebtedness cycle. He again took on lease the land from Bania for the similar amount of Rs. 1100 per year. This time he could hardly get 16 quintals of rice due to shortage of canal water/rains etc. This produce was sold to Bania for Rs. 1,470. Bania deducted Rs. 1,100 and paid Rs. 370 back.

Brahm Prakash was quite definite that in time to come, his debt burden would again increase. He did not take any loan from Mini Bank and other agencies because if Bania was not paid back his dues it did not matter so much. He could execute fresh pronote but if one defaulted to bank, one might be sent to jail as well.

Brahm Prakash felt, "I will prefer to take loan from Bania rather than Bank, no matter even at high rate of interest because this way I do not have to risk humiliation at the hands of Bank officials, in case of default. I know several cases in my village when several persons who took loan from Mini Bank were harassed even when some of them were genuinely hard-pressed this year because of drought. Some of the defaulters were taken into custody. Those who could manage the amount of the sureties were released, others remained in custody for 40 days. They were released only after executing the guarantee that they would repay the remaining amount totally at the time of next crop along with the interest and expenses involved in legal process".

Brahm Prakash narrated case of another farmer, Tara Singh Sardar who had taken loan of Rs. 3000 from Mini Bank. When the amount fell due he was asked to pay Rs. 3,600 including the interest. Tara Singh did not have much income due to drought. Because of the continuous harassment at the hands of Mini Bank officials he was forced to borrow Rs. 3,500 at the rate of interest of 5 per cent per month from moneylender to pay back the bank dues. Probably he would have been saved of this harassment and exorbitant interest rate had he borrowed only from Bania, Brahm Prakash felt.

He further mentioned that seeds and fertilizers given by the bank were not of good quality, neither were they provided well in time. Even the choice of fertilizers and seeds was imposed on farmers. The bags of seed and fertilizer were usually under-weight or of adulterated quality. Brahm Prakash when asked, how without being the member of bank he could say such things about its dealings, he replied "I carry the seed and fertilizer in my cart whenever fertilizers come to bank, even when farmers get it from the bank, they make use of my cart and therefore I know".

2. Shyam Chand

Profile

- —Family includes 10 members, including old parents and a younger brother studying in 10th class in the same village.
- —Owns a pucca house with three rooms, verandah, kitchen etc., electrified.
- —Main crops grown in kharif are paddy, bajra, jowar and chilly, in Rabi—wheat.
- —Shyam's father is a member of Cooperative Society (Mini Bank) since 1963 and reportedly never defaulted in repayment.

Case

Paddy being major kharif crop of his area, Shyam Chand had developed nursery for planting paddy in two acres. He had taken a loan from Cooperative Society to the extent of Rs. 1200 out of which Rs. 500 worth was given in form of fertilizer, rest being cash.

The paddy was transplanted in last week of June. The electricity supply was smooth and early setting of seedling was quite promising due to good soil conditions and water supply. But soon his spirit was dampened.

The monsoon totally failed and the electricity supply became erratic. There was no supply in day time and during night also, heavy voltage fluctuations were there. During one such night, his motor was seriously damaged and he had to spend about Rs. 250 for repairs. Power position was worsening and paddy crop starting showing symptoms of water stress. He suggested to his father that they should also go in for diesel engine. An amount of Rs. 2,500 was borrowed from a local moneylender @ 3 per cent per month interest (i.e., Rs. 36 per cent p.a.). With the help of a mechanic, he bought an engine worth Rs. 2,400 from Panipat. In all, it took him 10 days to get engine installed after incurring an expenditure of Rs. 200.

Paddy could not withstand such long spells of water stress. The wilting started. Non-availability of diesel added to Shyam Chand's problems. Several farmers joined hands in sending complaints to Agricultural Department, District officials, including S.D.M., Collector and even Chief Minister. When pressure increased, the government issued an ordinance instructing electricity department to supply power in day time and 60 per cent of the total state's supply was to be diverted to rural regions. Power position did not improve much. Farmers were told that neighbouring states were not giving Haryana its due share of electricity.

Government, in view of the persisting problem instructed District Collector to authorise Block Development Officers to issue permits for diesel to the farmer owning tractors and diesel engines.

After adoption of Training and Visit System of Agriculture extension, the B.D.O.s had lost direct contant with the farmer. They had to take help of Agriculture Department.

The criteria laid out for the distribution of diesel was that tractor owners would get 50 liters and those owning diesel engine would get 20 liters per week. To prepare the list of names of farmers owning either tractor or diesel engine, the Circle Agriculture Officer asked the Agricultural Development Officer, village extension worker and agricultural inspectors to prepare a list of the farmers of the village under their jurisdiction giving the required details. Shyam Chand mentioned it with a sense of bitterness that the list prepared by them was not accurate. Some of the farmers who did not have any of these things managed to get listed and in turn bogus permits were issued, while some owning tractor or engine in fact were left out. Shyam Chand was one of them.

He complained to the Circle Agricultural Officer (CAO) and produced before him the cash memo of the engine that he had recently purchased. CAO asked him to give application to this effect duly verified by Patwari and Sarpanch of his village. He complied with this within one day. CAO took his application and asked him to come next day to collect his permit. Next day when he went to his office to enquire about the same, he was told that his name was there in the list and he had already been issued the permit. CAO showed him the list where his name was mentioned on serial No. 19 having a thumb impression embossed against it and already certifying that the permit had been issued. Shyam Chand argued with CAO and tried to convince him that he had not got any permit. The CAO threatened him that complaint would be lodged with the police against him if he persisted. Shyam Chand couldn't say anything further on this. He came back, sad and disillusioned.

Next day he went to Panipat to complain to SDAO (Sub-Divisional Agricultural Officer) but in vain. In this process eight days were lost. Paddy crop was almost on the verge of complete failure.

He tried some how to purchase diesel in the black market and tried to save his crop. But it was too late. The paddy crop was lost. And with it, many of his hopes.

He was to pay back to the moneylender, the amount that he had borrowed to purchase diesel engine.

Bajra crop that he had grown largely for fodder purpose also dried up. There already was foodgrain shortage due to drought, the fodder problem forced him to take decision to sell off his buffalo. Shyam Chand was sure that in normal times, buffalo with about 10 liters milk yield per day, would have easily fetched him an amount of Rs. 2,500. The interest on the loan was piling up. He paid Rs. 1,800 to the moneylender and promised to pay back remaining Rs. 700 soon.

The loan from the Mini Bank could not be paid back due to drought and crop failure. He tried to get another loan from Bank for purchase of wheat seeds and fertilizer. Now that power condition had improved, he thought, he might be able to recoupe his position a bit. Mini Bank officials refused to give him another loan till he paid back the earlier one. He knew, many farmers, influential and powerful ones, had managed to get second loan without paying back the first.

He sowed wheat after borrowing seed from some other farmers but borrowing fertilizer would mean heavy interest. He couldn't apply enough fertilizer, to get maximum yields. He did'nt want to borrow more from moneylender. An amount of Rs. 700 and interest remained to be paid back, just like the Mini Bank's loan. If he did not pay back this time too, he could be put in jail. He didn't want to risk that but yield was to be low.

He had seen his earlier crop dying. He was witnessing the present crop too, with some sense of gloom.

3. Duli Chand

Profile

—Fifty years old educated up to primary class, two brothers living separately.

—1.5 acre land inherited from his father who died some years ago.

—Purchased 10 acres of land near G. T. Road in 1968.

—Two sons both married and in service living in joint family.

—Owns big pucca house consisting of 8 rooms and two verandahs, etc.

—Treasurer of cooperative society.

Case

Twenty-three years ago, Duli Chand enrolled 20 farmers as members of a cooperative society which was later registered. Duli Chand became the treasurer and his cousin, the President of the society. Other office bearers were also choosen from amongst his family members. Duli Chand remained treasurer from 1958 to 1976. During this period, through various dubious practices of his, he got some small farmers enrolled as members who were not even aware of it. Their membership fee etc. was deposited by him in their names. He used to draw loans every year in their names, and utilise the money for his own business. He also started lending money on higher rate of interest i.e. 3 to 6 per month. Not only this, he indulged in trading of foodgrains, Gur (Jaggary) and other commodities, stocked them and speculated in the market. He made lot of money in the process. Within 10 years he had enough money to purchase 10 acres of land worth about Rs. 80 thousands near the National Highway, (G.T. Road).

He managed the business of taking fictitious loans very cleverly. He would take loan in name of person 'A' and at the time of repayment would either get it renewed or through another loan in name of 'B' would close the loan taken in the name of 'A'. Through this "Naya Purana System" (New and Old) he ensured that these loans never became bad on records and therefore the farmer would never be issued a notice which could inform him of the reality.

He purchased after some years, a car which he used as a taxi. Soon the taxi was disposed of because of the increasing demand of money for drinking and other luxuries that he indulged in. After some time he purchased a motor cycle for his personal use.

The way he managed his affairs astonished the entire village. No body knew how could he get so much of income ! He had developed effective links with the influential officials in the higher hierarchy of cooperative department. When a complaint was lodged against him to the Registrar of the Cooperative Society in 1967, it was found that his records were quite upto-date and there was nothing improper. He continued with the practice of 'Naya Purana' even after the inquiry. The names of the farmers against whom he used to draw loans every year were never included in the list of defaulters required to be sent to Central Cooperative Bank for issue of demand notice because these fellows were never defaulters on records.

In 1976 all the Cooperative Societies in Haryana were dissolved due to managerial malpractices and accumulated losses. Government decided to re-organise these societies in the form of

Mini Banks. These Banks had the jurisdiction over a Patwar circle. While its management was still in the hands of elected representatives, its cash transactions were maintained by full time paid and trained secretary-cum-manager who was an employee of Central Cooperative Bank.

On review of records of this society, it was found that number of persons were defaulters including the secretary and the treasurer Duli Chand himself. The persons who were fictitious members were also found to be defaulters. The farmers whose names thus figured in the list were amazed to find themselves as defaulters while they had never enrolled themselves even as members. When notices were issued to them for a total sum of Rs. 42,000 (forty-two thousands), it was revealed that the list included a farmer who died before the date of borrowing loan from the society. This exposed the game of Duli Chand quite a lot.

One of the victims of this fraud, Maha Singh gave the information that in 1977, one day when he was working in the field, officials of Cooperative Bank came in a jeep and asked him to accompany them to the village. When they reached there he found the other victims also present. All the farmers informed the Officer that Duli Chand had managed to take loans in their names without their involvement at all. Officers immediately contacted Duli Chand who admitted his mistake. He was handed over to the Police where he was kept for 40 days. He was released on bail and a case was registered in court which is still pending.

Duli Chand had disposed of 5 acres out of his land to manipulate and block the enquiry. All the 8 victims had also filed the case against Duli Chand in civil court withdrew after he admitted his fault and promised them to repay the loan. The disputes continue while the victims suffer.

Conclusion

The three stories narrated above support some of the arguments we have been making about the impact of Green Revolution in the district. Imperfections in credit and product markets left very little choice to farmers like Brahm Prakash who grew paddy but obviously with not much gain. They can neither apply proper quantity and quality of inputs at the right time nor can they market their produce at the best time and place. Brahm Prakash was perhaps right when he said that his indebtedness cycle would continue. He had disposed of his only possession—half acre of land. He did not know what his produce would be next time but he was confident that the Bania would continue to lease land to him and advance him credit. The banks would not do either. The fear of repercussions of bank default was not perceived in reference to the Bania. Clearly institutional finance was unable to break his indebtedness cycle and therefore probably it was 'rational' for Brahm Prakash to choose not to participate in official programmes of credit and input distribution.

Shyam Chand goes through a similar frustrating experience with the bureaucratic authorities. He is much better off economically but is a victim of wrangling procedures and local power politics. He relies on the moneylender to ease his troubles and fears jail for bank default.

Duli Chand demonstrates the procedures that harm the interests of farmers like Shyam Chand and bring down the credibility of cooperative banks. For 23 years Duli Chand and his other accomplice had managed to deceive the farmers and cheat the Cooperative Society.

Much is made against the exploitative nature of the role of moneylender. What is forgotten is that he plays a role that finds easy acceptance among deficit budget farmers (Gupta, 1980) who are caught in an unending indebtedness cycle. The intervention of institutional finance is too heavy. Unable to break this cycle, the bureaucracy turns to those who carry, the capacity to utilize its interventions and hence satisfy the demands of fulfilling targets and goals. The new agricultural strategy was not interested in such groups of people and hence the bureaucratic implementation meant disbursement of loans to those who were the 'entrepreneurs'.

7

The Sample and the Social Background of the Respondents : Officials, Non-Officials and the Farmers

THE PREVIOUS chapter amply demonstrates the rapid strides Karnal has made in the area of agricultural production. This is reflected in increased agricultural production of various foodgrain crops, increased consumption of fertilizers, greater use of high yielding variety seeds and a remarkable expansion in development has been distributed unevenly.

We have concentrated our attention on two blocks of Nilokheri and Samalkha for further investigation. These two blocks sharply bring out the way development has taken place. Nilokheri had shot into national prominence with its experiments in Community Development by S.K. Dey who made it a training ground for many batches of extension workers. Population, wise it is a larger block than Samalkha (see Table (7.1). Both the block headquarters fall on the national highway that bifurcates the district. The net area cultivated in Nilokheri is 38,710 ha. and 19,818 ha. in Samalkha. Even though the proportion of net area irrigated to net area sown is similar in the two blocks there is a difference in the sources of irrigation. Irrigation by private tubewells is much higher in Nilokheri than in Samalkha while canals irrigate nearly the same area in the two blocks. Nilokheri has much higher proportion of area under each crop and also consumes much higher quantity of fertilizer per hectare. Of the total number of tractors in the district, it has thrice as many as in Samalkha. Similar is the case with threshers. Looked at in another way, there are 19 tractors per thousand hectares cultivated in Nilokheri and 13 for the same area in Samalkha. There are in the same point of view 33 threshers per thousand hectares in Nilokheri and 21 in Samalkha.

TABLE 7.1 : **Block profiles : Nilokheri and Samalkha**

Characteristics	Nilokheri	Samalkha
Number of culivators as % of total workers	50.34	51.74
Number of agricultural labourers as % of total workers	29.28	19.64
Net cultivated area in hectares	38710	19818
Percentage of net area irrigated to net cultivated area	87.77	89.84
Percentage of area irrigated by source to net cultivated area :		
State Tubewells	0.60	2.58
Private Tubewells	61.49	55.38
Canals	25.68	27.40
Percentage area under crop as % of gross cropped area :		
Wheat	47.59	10.09
Paddy	49.57	9.88
Maize	4.39	2.38
Average quantity of fertilizers used (in kg.) per ha. of gross cropped area	198	107
Number of tractors per 1000 ha. of net cultivated area	19	13
Number of pump-sets per 1000 ha. of net cultivated area	205	144
Number of power threshers etc. per 1000 ha. of net cultivated area	33	21

Source : District Credit Plan, 1980-82, Punjab National Bank, New Delhi.

Obviously, agriculture is more mechanized in Nilokheri than in Samalkha. Probably, a consequence of such mechanization is that the proportion of agricultural workers to total workers is much higher in Nilokheri than in Samalkha even though the proportion of cultivators is nearly the same in the two blocks. These are figures of the 1971 Census and there may be changes now. But considering the 1977-78 figures on mechanization the trend may be undisturbed.

A further feature regarding banking facilities may also be mentioned here. There were total number of 108 branches of commercial banks, cooperative banks and land development banks in the district. In Nilokheri block there were 12 such branches and only 7 in Samalkha. In addition, while branches in Nilokheri were fairly spread in the block with only 3 branches in Nilokheri proper, the picture in Samalkha was quite different. Out of 7 branches in the block, 4 were located in Samalkha itself. (see District Credit Plan, Chart opp. p. 36). In passing it may also be interesting to note that among the schemes sanctioned by Agricultural Refinance and Development Corporation in the district, no scheme is located in Samalkha block while Nilokheri is included.

The sample of respondents

The sample of our respondents has been chosen from these two blocks, Out of our respondents, 31 are officials, 29 are non-officials and 100 are farmers. They are distributed in the two blocks as shown in the Table 7.2 below.

TABLE 7.2 : **Distribution of the sample district Karnal and, blocks Nilokheri and Samalkha**

	Karnal (district level)	Nilokheri	Samalkha	Total
Officials	8	10	13	31
Non-officials	—	12	17	29
Farmers	—	50	50	100

The officals and the non-officials

Out of the 31 officials in the sample, five of them were functionaries at the District level. Fourteen of them were village level Agricultural Development Officers. The rest were located at the Circle/ Sub-division level. All the 29 non-officials were Sarpanch i.e., elected heads of the Panchayats at the village level, The average age of the officials was 37.4 years while that of non-officials was 50.8 years. There were at least 60 per cant of political leaders who were above the age of 45 years while among the officials the percentage was only around I3 per cent. Only 58 per cent of the officials but all the non-officials were born in the State of Haryana. There is also another difference between the two. As expected, the officials were far more formally educated that the non-officiais. Around 80 per cent of the political leaders were educated up to primary level while a similar proportion of officials were college graduates.

Despite some of these different characteristics of the two groups, they appear to come from a very similar farming background. At least 90 per cent of the officials were born and bred in agricultural families who still continue their farming operations, while they themselves have moved on to jobs in the government. All the political leaders are farmers themselves. Thus, while the factors of age, education and profession tend to separate the officials and the non-officials, the two groups appear to be similar in terms of their family background. Both come from farming families and while the political leaders are agriculturists even today, the officials go to their families to render help whenever time from their other duties permits.

The farmers

A random sample of 100 cultivators was chosen from 524 farmers residing in four villages of the sample blocks. Table 7.3 shows how they were distributed according to their size of holdings. As can be seen, 42 per cent of the farmers owned land up to 2 hectares.

TABLE 7.3 : **Farmers in the sample according to size of land holdings**

Size of land holdings in hectares (ha.)	Number in the four villages	Nilokheri	Samalkha	Total
0.0 — 1	90	8	13	21
1 — 2	134	9	12	21
2 — 4	214	14	17	31
and above 4	66	19	8	27
Total	524	50	50	100

In terms of sources of irrigation, there are 75 per cent farmers in the sample who have their own tubewells. Rest of them depend on canal irrigation and use tubewell of their neighbours through payment. Table 7.4 depicts the source of irrigation according to size of land holdings.

TABLE 7.4 : **Sources of irrigation according to size of land holdings of sample farmers in Nilokheri and Samalkha**

Size of land holdings in (ha.)	Samalkha		Nilokheri		Total
	Own Electric/diesel	Hire/Canal	Own Electric	Hire/Canal	
Up to 1	3	10	—	8	21
1—2	8	4	9	—	21
2—4	16	1	13	1	31
and above 4	8	—	18	1	27
Total	35	15	40	10	100

It is quite obvious that tubewell irrigation increases with the size of land holdings in both the blocks. Larger farmers in both the blocks hardly utilize canal irrigation. Out of 75 cultivators who own their own diesel/electric pump-sets, only about 27 per cent have less than 2 ha. of land holdings. Proportion of farmers having their own sources of irrigation by owning diesel or electric pumping-sets is higher in Nilokheri than in Samalkha.

A major consequence of the pattern of irrigation can be seen in terms of the cropping pattern of the sample farmers. Paddy growing appears to be closely linked with the size of land holdings and tubewells as sources of irrigation. See Table 7.5.

TABLE 7.5 : **Cropping pattern according to size of land holding in Nilokheri and Samalkha**

Size of land holdings (ha)	(Area in ha.) Nilokheri			(Area in ha.) Samalkha		
	Paddy	Wheat	Maize	Paddy	Wheat	Maize
Up to 1	6.48	8.10	—	0.20	11.5	40.61
1—2	12.96	14.98	—	1.42	15.79	0.20
2—4	32.59	40.49	1.01	8.91	42.91	1.82
and above 4	167.61	150.21	2.43	45.26	42.11	2.83

A consequence of ensuing prosperity is reflected in terms of education of the farmers. Forty-two per cent of the total farmers in the sample are literate but among the literates about 64.6 per cent of them belong to Nilokheri. Again among the literate farmers (n=42), see Table 7.6, about 65 per cent of them are those whose operational holdings are of more than 2 ha. Similarly, out of these who have been educated up to high schol (n=24) about 70 per cent of them belong to the same land size category. Apparently, the incidence of education goes up with the size of land holdings. Thus in case of cultivators in the sample, the size of land holdings is a significant variable determining the ability to gain education.

TABLE 7.6 : **Educated farmers according to size of land holdings**

Size of land holdings	Level of Education			Total
	Primary	Middle	High School & above	
Up to 1	2	1	2	5
1—2	2	3	5	10
2—4	0	4	7	11
and above 4	0	6	10	16
Total	4	14	24	42

Conclusion

The data on the sample tends to support the policies that were adopted to implement the green revolution strategy. There is a remarkable difference between the two blocks of Nilokheri and Samalkha because of the fact that predominant source of irrigation is different. Other technological inputs appear to be related to this variable—even to the extent that Samalkha has not been able to attract financial institutions in the same way as Nilokheri. Thus, the opportunity for a well endowed area to prosper goes lower than a district creating disparities among blocks and probably among villages too. The strategy of "betting on the best" is fairly obvious even below the district level.

Further, the major or crucial variable affecting the life of a cultivator is the size of his operational holdings. Larger farmers are able to attract several facilities. There were 58 per cent of the farmers in the sample who owned more than 2 hectares of land. Out of them, 96.5 per cent owned electric or diesel pumping-sets for irrigation. It is these farmers who have taken to growing rice. Thus, it is this class of farmers who own comparatively larger operational holdings, own tubewells for irrigation and who are paddy growers in our study. They have also taken to paddy growing because it is a commercial crop and most of it is exported.

Another consequence of such prosperity is on the educational pattern of our respondents. Among those who are educated beyond high school, more than half of them are those who possess at least more than 2 hectares of land. This leads us to an obvious and well-known conclusion that education and larger land holdings are positively related. With this fact, another finding that needs to be mentioned is that the officials come from farming families and continue to carry on the agricultural interests while performing their jobs as government functionaries. This means that due to requirement of education among government officials, they must be coming from families that must be owning more than 4 ha. of land. Thus, the rural background of bureaucrats has to be recognised as that of a comparatively large landowner. Such a close definition of rural background becomes significant in a framework when the government's objectives in promoting rural development are to help the rural poor. Even though the background of the bureaucrats is rural, it may not be in line with governmental objectives because of the narrow social base. Hence, more often than not, one finds in actual experience, that those coming from the rural background in the administrative system tend to be more negative towards certain developmental policies than those from urban background. Probably, this happens because the socialization of the rural bureaucrat is that of a rich farmer and not of a landless labourer, sharecropper or a marginal farmer. What thus happens is that widening of the educational base or recruitment pattern, leads to the grabbing of opportunities by those who are better off in the rural areas. They are accepted as representatives of the rural orientation but may, in all probability, not act in sympathy with those who are deprived. They may project biases and prejudices that emanate from the fact of being better off.

Thus, it would be misleading to conclude that because the local functionaries have a rural background, they will perform better as development administrators. For, if development is defined in terms of changes in the life of the rural poor, there may be greater chances of such a bureaucrat not being in sympathy with these objectives. For achievement of these objectives will directly bear on the pattern of life his own family leads and that of the environment in which he has been brought up.

8

The Administrative System and Development

IN THE preceding chapters, we have attempted to highlight the implications of the national policies that were adopted to bring about Green Revolution and the administrative context in which they were implemented. The ensuing chapter is concerned with the way the local level administrators, political leaders and the farmers view the changes that have been taking place around them. The purpose is to investigate how they feel about the way the interacting components of the system have developed and have affected them. In trying to investigate the perceptions of the administrators and local political leaders of the working of the administrative system, we asked them the following questions :

1. All of us want our government to do certain things. The success of a government in doing these things depends on how it is run—administered. When you look at administration in Karnal district/Haryana, how do you feel about it ? Please describe.
2. Now taking the opposite view, what are your fears and worries about administration ? What are the worse things that could happen to it ?
3. Here is a picture of a ladder. The top 10th step represents the best possible administration in Haryana and the bottom 1st step represents the worst possible. Now where would you place the administration of Haryana—five years ago, now and five years hence ?

4. What do you think should be done to improve the administration ?

Among the officials, three dominant characteristics of the best possible administration have emerged. First one is that of coordination. Most officials have lamented the fact that the developmental programmes are poorly implemented and coordinated. As one of the respondents put it "every officer should see the village at least once a month. Only then will he be able to find out how the developmental programmes are being implemented". It is in this connection, another respondent has mentioned that the administration under a former Chief Minister during the days of national emergency was much better. There was clear line of command and programmes did not suffer due to lack of coordination.

The second characteristic has been expressed in terms of honesty. Most officials look at the present administration as "loose". They find administration as lacking in firmness and sheltering dishonest officers.

The third characteristic as mentioned by the officials is that of instability in government. The time the interviews were taking place, Haryana was going through a period of political change. Defections from one party to another were common and the State was plunged into political instability. Hence, there is some concern about this factor influencing the administration of the State. For, those who have elaborated on this theme have attempted to establish a relationship of the quality of the administrative system with that of the political leaders. As one respondent explained, "administration of an organization depends on the machinery that governs it. If that is occupied by well and noble persons who are responsible and are concerned about timely implementation of programmes, then administration will be good".

Finally, there was one official who expressed concern about developmental schemes not being able to reach the intended beneficiaries. He thought that, for example, giving loan or credit to the deprived sections of the society should not be a time-bound or target fixed scheme. When the targets are fulfilled under pressure of time, the administrators are more concerned about achieving paper results. Consequently, the intended beneficiaries are neglected and those people get included who do not fulfil the necessary requirements.

Among the local political leaders, the response to the first question is dominated by expressions of concern about corruption. Most of them feel that corruption has increased and has to be reduced considerably in order to improve the administrative system. Some have found corruption as a consequence of political instability. One of the respondents felt that "administrative structure at the State level is not good. This had led to the increase of corruption at the district level". He then went on to assert that "I think people of this country do not deserve all such facilities and freedom. They must be governed strictly". Such a relationship between rising corruption and declining discipline and strictness has been mentioned by several of the political leaders interviewed. Many thought that the administration during Emergency was much better.

One of the political leaders sought to look at corruption as manifestation of distortions in distributive justice. He felt that schemes meant to help the poor tend to get diverted towards the rich and thus vitiated rural development. Increasing corruption meant in one way the growing neglect of the poor.

In responding to the second question, the fear about the future was dominated by the thoughts of rising corruption among the officials. Most thought that corruption will increase. For some it meant, "no further development will take place". An official thought that the progress of development programmes depends on the efficacy of administrative machinery. If it declines, it will have direct impact on what happens in the area of development. For others, increase of corruption will mean poor standards of administrative efficiency for administrators will lose all interest in their work. Promotions or dismissals will be arbitrary and the administrators will lose faith in the way the machinery operates.

A respondent also expressed a fear about the impact of corruption on agricultural production. He felt that the Green Revolution was sustained through heavy doses of inputs like water, chemical fertilizers, improved varieties of seeds etc. If they were not make available to farmers on a timely basis, production will suffer. He thought that increasing corruption will have an impact on the timeliness of delivery of inputs.

For the political leaders, future was beset with rising corruption. Many felt that this would lead to a serious loss of government's credibility and reputation. They felt that this would result in complete failure on the developmental front.

Among the farmers too the dominant theme in reference to government and administration is that of corruption. Most of them have mentioned that the administration is dishonest and the administrators work in their own self-interest only. "Nobody cares for us" is a common refrain. Many of these farmers have related the increase in level of corruption to instability in government. One of them remarked, "government itself is busy in establishing itself. It has no time to pay any attention to administration". Similarly, another farmer has added that, "the government is so busy in saving its own position that it has little time to devote to tasks of development". Some have also joined this refrain by comparing the present government with the one prevailing during the time of National Emergency. This comparison is used to mention lack of progress, difficulties in getting inputs and rise in prices of essential goods. A farmer has summed up this situation by saying, "at present government and administration is composed of selfish persons. They are doing each and everything in their self-interest".

A second dominant theme is related to the distributive aspects of benefits flowing out of development schemes. One of the farmers has put it candidly when he said, "if you like to know whether the government is good or bad you should meet someone else. . . I would only like to say that government is nothing but a group of rich people and whatever they do or are doing are only for the rich people. . . We poor people get very little". Another respondent said, "the character of the government has fallen down. Nothing is available in time for the poor. But, on the other hand, the rich and the influential persons are now in a much better condition". Thus, there appears to exist a strong feeling that the poor have not reaped many advantages of development. These have been cornered by the rich and the influential.

Farmers have expressed their fears of the future primarily in terms of increase of corruption among the government functionaries. The feeling is that the growth of corruption will lead to complete distortion and stoppage of development schemes. One of them has pointed that "we may be worse off than what we were before independence".

Thus, while around 30 per cent of the farmers who responded (73 per cent replied to these questions) did not find much fault in the administration, the rest found falling standards in administrative morality and increasing disparities between the rich and the poor as the major characteristics of the prevailing administration. What the farmers feared most was increasing level of corruption.

These descriptions have been assessed by evaluating the administration on a ten-point scale. The average evaluation by the respondents are presented in the Table below :

TABLE 8.1 : **Mean ladder ranking**

	Past	Present	Future
Officials	8.4	5.8	6.8
Political Leaders	7.4	6.8	5.4
Farmers	7.6	6.0	6.2

From the above, it is evident that the average ladder ratings by the three groups went down from the past to the present. However, the ratings of the past differed. While around 69 per cent of political leaders and 53 per cent of officials rated past higher than the average step, only 27 per cent of the farmers rated it so. Therefore, while it is true that past was evaluated better than the present, this feeling was weaker among the farmers than the other two groups.

In comparing the present with the future, the officials were different from the other two groups. The average step ranking rises in the case of officials but declines for both the farmers as well as the political leaders. Moreover, there were around 47.6 per cent of the officials, a little less than half, who rated the future below the average step. Among the political leaders the proportion was only around 27 per cent while among the farmers it was highest among the three groups i.e. around 65.5 per cent. Thus, it may be safely concluded that in general there is a similar trend of pessimism among all the three groups but is marked by different intensity.

Among the farmers, there are significantly more farmers owning more than 2 hectares of land who do not think that there will be any change between the present and the future. But the same group feels the present has slided down from the past. (Chi-square in first case was equal to 10.6 and in the second to 8.9). Thus, the farmers in this category feel that though the past was better: future may not be worse than the present. This assessment of the past and the future must be looked at within the suggestions of a few in this group that; (a) the emergency rule way much better, and (b) the rule of Bansi Lal as Chief Minister was a great contribution to the development of Haryana.

In terms of suggesting improvements in the administrative system, there appears to be a general agreement among both the officials as well as the political leaders that the administration should be strengthened. It must have the capacity to punish the guilty and implement the policies of the government without fear or favour. But the process of strengthening administration was perceived very differently by the two groups. The officials were primarily concerned with the reduction of what they called political interference in administration so that, as one respondent has pointed out, "power of the officers (is) properly and rightly utilized. The only way to improve administration is to give full freedom to the officers". Another respondent has elaborated this point by suggesting, "that jobs should be specifically assigned to officers at various levels and once that is done, they should be free to work. All sections of society may be associated with the review of this work but till then the officers should be left alone". Another respondent has added that freedom to work includes framing of rules and regulations necessary for the completion of a job. The official may be permitted to do so because he is the only person who is best acquainted with administrative constraints and local conditions".

A few of the respondents have suggested that the administration can be improved only if there is a consensus between the higher officials and political leaders. If there is disagreement at the top, it is bound to percolate to the lower levels of the administrative set up and it is here that greater difficulties crop up.

The political leaders, have an entirely different perception of the process of strengthening or improving the administrative system. Most of them feel that the prevailing administration is only a reflection of the character of political leadership and government stability. Strong leadership and stable government were a prerequisite of effective administration. Their emphasis is on political leaders "who should be strong enough to take decisions", or "who have adequate discipline themselves so that they can take action against the wrong doers". For such leadership or stability, most of these respondents suggested that educational system must be invigorated so that the political awareness of the masses rises and they do not remain apathetic about the activities taking place in political and economic spheres and thus exercise effective control over them. They lamented the fact that more and more people had become selfish and were concerned about themselves. Such a situation could be remedied by education and increased political awareness.

A few respondents in both the groups have mentioned the level of patriotism in the people as a crucial factor impinging upon the administrative processes. They feel that it is necessary also "to create a feeling of patriotism among the bureaucrats and the political leaders" before administration can be improved.

Officials have also given some specific suggestions to improve the administrative system. One of them would like to see much greater decentralization in decision making for the local level both in the spheres of finance and general questions of promotions and transfers. Another respondent would like the process of work to be redesigned so that files can move up and down at greater speed. A few of the officials have commented on the relationship of the superiors and subordinates in administration. They feel that the subordinates are usually ignorant of the aims and objectives of a policy and are informed only of what is to be implemented, not why. Finally, a respondent has felt that the superiors must spend more time in guiding and supervising the subordinates. This can be done only if they had greater interest in their welfare.

The political leaders did not have much to say on the specific suggestions to improve the administration. However, it is interesting to point out that one of them suggested that to raise the level of honesty among administrators, they should not be posted to the same district or village to which they belong.

Summary

To conclude, it is significant to point out that a major theme that has emerged out of the perceptions of the three categories of respondents of the administrative system is that of honesty and corruption. Whichever way one looks at administration, the general feeling is that it suffers from lack of integrity at both administrative and political levels. However, there is much stronger acceptance of the fact that administration is merely a reflection of the character of political leadership and government stability among the political leaders than among the officials. Hence, strikingly suggestions of improvement follow a different pattern. There is a greater emphasis on general societal progress through education by the political groups than the officials. However, a general feeling of pessimism seems to be shared by all though the strength of this feeling differs among the groups. The farmers appear to be more confident than the other two groups while the officials seem to be least so.

Another interesting feature of these perceptions is that the farmers are the least pessimistic about the future while the administrators in comparison are the most so. This probably is a reflection of the Green Revolution environment on the farmers but not fully shared by the government functionaries.

It may be reasonable to conclude on the basis of several other studies (Mathur, 1972, Singh & Ashraf, 1971) that administrative perception of pessimism and frustration are a wide ranging characteristic whether the administrators are located in an area undergoing rapid rate of development or not.

Apparently, it is the gap between what has been done and what can be done that raises the kinds of dissatisfaction that have been mentioned in the earlier sections. There is widespread feeling that the cause of this gap is corruption at both the levels of administrative as well as political leadership. Belief in the pervasiveness of corruption is a striking characteristic of the decades of planning in India. Everyone accuses everyone else of a style of living and behaviour without in the least realizing that they themselves may be indulging in the same. Highest of moral and personal qualities are used as standards to measure others but not oneself. The result is a constant stinging match in which it is very difficult to discern who is really corrupt. The "holier than thou" attitude pervades all walks of life and all people perceive corruption to be prevalent at levels other than their own. It is within this social environment that we have to place the dominant perceived qualities of good administrators, good political leaders and the administrative system as a whole.

Another characteristic of the planning decades has been that development is viewed as the responsibility of the official agencies and not of the people as such. Any shortfalls in production or distribution are attributed to the failures of the government and the administrative system. The result is that everybody associated with such official agencies finds himself within the general syndrome of accusations of corruption and lack of integrity because failures on the developmental front are blamed on him.

A consequence of such types of social perception is that the job skills and qualifications or the ability to perform a role do not get the requisite importance in the perceived qualities of a good administrator or a political leader. Similarly, a good or bad administrative system is also perceived in moral terms. Very few of the respondents, and that also among the non-officials, for example, would like to see greater knowledge of agricultural practices among the government functionaries as related to the farmers. It is only a few officials who would like to judge the political leaders from what they can do for the constituents or the administrators from the speed and efficiency in decision-making.

Among respondents who are cultivators, there is a mention of an important feature of the administrative system that does not find a place among the perceptions of either administrators or the political leaders. This relates to the ability of the government to perform in the sphere of distribution of the inputs necessary for increased production. It is this group that feels that the rich are the more favoured among them and the government and its administration should be judged on the criteria of their capability to help the poor. Their criticism within this broad theme is quite strong in some cases. As is to be expected, such a dimension of thinking emerges more strongly among those farmers who own land up to 2 hectares. It is this group that also tends to associate corruption with cornering of benefits by the rich. Comparatively, and, by and large, the farmers in larger categories of size of landholdings perceive non-availability of inputs in quantity or in time as a sign of corruption. Therefore, corruption even though a widely prevailing perception, does not have uniform content for all farmers.

Thus, the theme about corruption and honesty has three distinct components. One is related to timeliness of input delivery. As we have already explained earlier, the agricultural strategy being employed is in terms of a package where not only are the inputs important but also the time at which they are available and made use of. Thus, availability is only one part of the package deal; the other is timeliness. Hence the significance of linking of corruption with not getting inputs on time.

The second component is concerned with aspects of distributive justice. The smaller and the poorer farmers find it difficult to explain the rising phenomenon of the richer and larger farmers gaining so many advantages. Unable to examine the implications of the strategy, corruption becomes a plausible explanation. The political rhetoric supports them in this view too.

It is only the third component that links corruption with lack of discipline. For many of our respondents mostly among the non-officials and the farmers, Emergency of 1975-77 holds the answer. The emphasis is on discipline and the capacity to enforce it up to the local levels. In fact, many have been mentioned the name of the Chief Minister who could provide an atmosphere of getting things done even without the declaration of the Emergency.

Lastly, it may be emphasized that while the creation of technical and infrastructural facilities appears to be symbols of successful development efforts, difficulties regarding availability of agricultural inputs or their diversion to those who are well off appear to be source of dissatisfaction and of assessment of corruption prevailing in the overall system.

9

The Officials and the Non-Officials

SINCE the introduction of Panchayati Raj and democratic decentralization considerable interest has been shown in investigating the relationship between officials and non-officials and their role in formulating and implementing developmental programmes at the local level. It has been argued that friction between the officials and non-officials is the dominant feature of their relationship and this has led to their inability to work as a homogenous team (Bhambhri, 1967; Iqbal Narain, 1966, Chaturvedi, 1964). The elected people have attempted to influence the officials to maximize the benefits for their own groups or their own areas. In these efforts they have been helped by their linkages with state level leadership and their membership of the ruling party. The administrators, on their part, have resented the injection of this kind of leadership and have perceived them as threats to their own power and influence (Mathur, 1972). This mutual hostility has often resulted in lack of concern about achievement of targets and inability to relate programmes to the good of all the people.

In another study during scarcity conditions in an Indian state where relief funding was available in ample measure, it was found that non-officials duly acknowledged the role of the administrators who in turn accepted the role of the politician in implementing the scarcity works programme. The administrators particularly emphasized the role of politicians in resolving disputes that related to land acquisition for scarcity works and their site selection. Overall, there was a positive reaction to each other (Mathur, 1974 : 9).

It is of interest, to ascertain the direction this relationship has taken in Karnal.

As we have already pointed out in an earlier chapter, there appears to be a vast difference of age and education between the officials and the non-officials. The officials are considerably younger and more educated than the non-officials. However, in terms of social background, the similarity between the two was remarkable. Invariably, both the groups belonged to farming families—while the officials returned to their farms to help out their families whenever time permitted, the non-officials still pursued the same occupation.

It is interesting to note that both the groups believe that lot of agricultural progress has taken place in Karnal. From their point of view, the factors responsible for increased agricultural production were the following as shown in Table 9.1. They have been ranked according to the frequency of responses.

TABLE 9.1 : **Factors responsible for increased rice production in Karnal as ranked by officials and non-officials**

Factors	Officials	Non-officials
Availability of irrigation facilities	I	II
Fertility of land	II	I
Farmers response to New techniques	III	III

As is apparent from the above, irrigation and fertility of land are mentioned as the two most important factors. The officials have however, given greater importance to the provision of irrigation. Both the groups have also given importance to the response of the farmer as a factor for increased rice production in Karnal.

Further, both the groups of officials and non-officials overwhelmingly expressed the opinion that this record of production could be improved still further. They have pointed out several ways of doing it as shown in Table 9.2.

While both the groups place adequate supply of inputs as the first most important method, other ranks are quite different. But the methods mentioned are linked to each other. For example, increasing the area under high yielding varieties of seeds demands increased irrigation facilities.

TABLE 9.2 : **Methods of improving rice production in Karnal as ranked by officials and non-officials**

Methods	Officials	Non-officials
Increasing the supply of inputs, HYV seeds, fertilizers, pesticides etc.	I	I
Increasing area under HYV	II	—
Better administrative coordination	III	—
Better extension services	—	I
Increased irrigation facilities	—	III

Therefore, one can say that by and large the views of the officials and the non-officials regarding ways to improve rice production are quite similar.

Despite these similarities, officials and non-officials do not think very highly of each other. Each group thought that the other did not take adequate interest in the implementation of development programmes. As one can see from Table 9.3 that at least 48 per cent of the officials thought that non-officials took little interest in development at the local level. Twenty-seven per cent of the non-officials had similar views about the officials. Obviously, in their own minds, each group thinks that the other does not perform its tasks well.

TABLE 9.3 : **How far do you feel local officials/non-officials are generally interested in the implementation of development programmes ?**

	Officials			Non-officials		
	Much	So-so	Little	Much	So-so	Little
Officials N=31		—		6.45	41.94	48.39
Non-officials N=29	3.45	68.97	27.59		—	

Such a perception is further supported by another dimension of their thinking. Both the groups feel there is little cooperation between them. This feeling is much stronger among the non-officials than officials. Probably, this reflects upon the views that most officials have that they are burdened by requests from non-officials for special favours. Few non-officials feel that officials place any demands on them. Thus, the situation is that it is the non-officials who are dependent on the officials to get their things done. Consequently, the image, on the one hand, is determined by the pattern of demands being placed and on the other hand, on the extent to which they are met. Both do not lead to a favourable relationship.

What is interesting is that both these groups feel that farmers in general do not take active interest in developmental programmes. This feeling is much stronger among the non-officials than the officials.

Finally, we asked our sample of officials and non-officials the following two questions :

1. What qualities do you admire most in a government servant ?
2. What qualities would you like to see in a good political leader ?

In talking about the qualities of a good administrator, the respondents appear to be talking about four categories of qualities. The first category envisages those qualities that are personal like honesty, morality, punctuality, politeness etc. The second is concerned with abilities related to decision-making viz., "he should have the capacity to take quick decision", "he should be a firm decision-maker", "he should be a strong administrator", etc. Closely associated with this is the third category in which such qualities like education, requisite qualifications, knowledge of agricultural practices are included. Finally, the fourth category consists of relational qualities like ability to get along with people, ability to deal with farmers properly, etc.

Using the above categories, the responses of officials and political leaders were ranked and are presented in Table 9.4. As can be seen from Table 9.4, overwhelming number of respondents feel that the most desirable quality among administrators is that of morality and honesty. All the three groups are unanimous on this count. Having achieved near unanimity here, however, they diverge in ranking other desirable qualities. For the political leaders, knowledge of the job and requisite qualifications for the job are next most important attributes of a good administrator. As one of them has pointed out, "he should have knowledge about agricultural practices; more than what the illiterate farmers possess". The officials have ranked this group of qualities as third while giving second place to abilities of decision-making. For both the groups, inter-personal skills do not have very high priority. This is a little surprising because in developmental activities these skills play a very important role. Administrators should possess them in strong measure because they have to cajole and persuade while the political leaders are affected by them because they interact with the administrators most often.

TABLE 9.4 : **Qualities of a good administrator**

	Officials	Non-officials
Personal qualities of honesty and morality	I	I
Decision-making ability	II	IV
Competence—qualifications and knowledge	III	II
Inter-personal relations	IV	III

The overall assessment of administrators is in the framework of morality and honesty by both the groups.

It is in the same framework that political leaders have been perceived also by the two groups. Using similar categorization, the responses are presented in the Table 9.5 below :

TABLE 9.5 : **Qualities of a good political leader**

(*in per cent*)

	Personal qualities, honesty, etc.	Service to the people	Ability to act and help
Officials	I	II	III
Political leaders	I	II	III

As can be seen from the data above, both the groups agree more or less in equal measure on the most desirable qualities of a political leader viz. that of honesty and morality. For the second and third category even though the ranking is the same there is a tremendous difference in the strength of desirability. There are many more officials who would like to see the political leaders being selfless and possessing the ability to act on behalf of the people than the political leaders themselves. As one of the administrators put it, "a true leader is one who can do something for the poor and forget his own self-interest". Another respondent supports this idea by mentioning that "a leader should be one who gets the administrators to work in favour of the poor".

The overwhelming desirable image of a political leader or a good administrator is moulded in a frame of moral and personal qualities. This concern for spiritual, moral and behavioural values seem to permeate the entire society. Several other studies have also found this the dominant predisposition in assessing people and events in society. Probably, a source of this orientation is due to the fact that appeals to dedication, public spirit and national conscience are constantly made and the perceived decline in society is blamed on the neglect of moral and religious principles.

Discussion

Two distinct dimensions of relationships between officials and non-officials appear to emerge from the data presented above. One is a broad agreement on the progress of agricultural production, potential of improvement and the ways of exploiting this potential. The other is that of difficulties in relationship due to poor image of each other. These difficulties largely emerge because of the dependence of non-officials on the officials for getting things done. Another factor influencing this image is the mutual evaluation of each others qualities. Both have a moral framework and do not rate the other highly.

Thus, probably the tension that may be ensuing is not due to disagreements on broad goals and specific activities taken up for increasing agricultural production but due to certain specific constraints. In addition whereas inputs are scarce, efforts must be to direct as much as of these scarce resources towards oneself. This process, using political or administrative linkages, puts pressure on the local level officials and possibly alienates them from both the political leaders as well as their own superiors. What is significant is that the tension is not based on policy disagreements but on personal predilections and attempts to divert advantages of development in a particular direction.

10

Bureaucracy and Policy Implementation : Some Tentative Conclusions

GIVEN the emphasis on technology and selective approach of choosing the area and the farmer for its application, certain compulsions in programme implementation emerged. These compulsions greatly influenced the scope and content of bureaucratic action at the local level. With little concern at higher levels for feedback on social implications, the local level bureaucracy was confined to fulfilling targets. As one of the administrators in the study pointed out, the targets being time-bound, they could be achieved more easily when the concern was not directed towards the group that was deprived but towards the group that had the capacity to utilize the inputs. Thus, technological achievement was symbolic of the administrative achievement too and therefore the measurement of success was in terms of number of pump-sets installed, amount of loan distributed, increase in acreage of paddy etc. The reduction of the planning process to the formulation of quantitative targets broken according to areas looked after by the officials moulded the bureaucratic role to concern itself with the technological achievements alone. To indicate, for example, to the district, block or village level development worker that the supply of fertilizer during the coming season will be roughly of a certain order is one thing. But to lay down to these functionaries targets for fertilizer consumption, as if it is they, not the farmers who use the fertilizers, is a different thing altogether. The bureaucratic machinery was quick to accept this type of challenge but did not or was not enjoined to control its social fall out.

Emphasis on the assumption of new technology being the harbinger of green revolution was also reflected in the way the Government responded to the needs of policy implementation. Administrative and bureaucratic expansion took place to provide inputs to the farmers. New technical departments or semi-autonomous organisations concerned with marketing, ware-housing, minor irrigation etc. were established. Credit brought in its own agencies. Each of the above organizations were asked to perform its task at the district level and below. Administratively this has meant the emergence of several parallel hierarchies at the local level. What started happening was that while the older hierarchies remained, new were added that nibbled away functions performed by the old multipurpose hierarchy. As we have seen in an earlier chapter, the team of development functionaries have continued but with limited functions. The T & V system has taken away most VLWs, and the agricultural department has initiated its own hierarchy with circles standing for blocks in their division. In addition semi-autonomous state corporations have branched out in the districts to provide inputs, facilitate marketing or help in storage etc. One agency may promote the use of an input, it is only another that will supply it and yet another that will establish the targets of consumption. Thus, the block development officer who concerned himself with area development has now been left with the work of Panchayati Raj and it is hoped that he will somehow get support from other departments in this vital task. As the targets were established for each of the tasks to be performed, coordination arose as the major problem in implementation. But so long as the larger farmers could pull in resources to satisfy their needs, the urgency for a coordinated effort did not arise. Those who suffered from lack of coordination neither had sufficient power to pressurise the administration nor stood as important partners in the green revolution strategy as far as the policy makers were concerned.

Coordination has come to be seen as a personal problem. The District Collector convenes monthly meetings of district level functionaries to motivate and to cajole them to help each other in achieving district targets. In such a situation, bureaucratic rivalries, departmental jealousies, political influence have a free play. It is the outcome of such dealings that leads to what actually is implemented or achieved. Another way in which the bureaucracy is open to outside influences in through the role of local level elite. Expanding employment opportunities in the bureaucracy brought in its fold such people who had background of agriculture. But this background was of a particular kind. The ability to gain education was determined by the size of the land-holdings. Farmers who had large land-holdings, and could spare their extra hands, educated their sons and relatives and enabled them to join government. Such bureaucrats working in extension agencies continue to spend time on their family lands on a casual basis and give them the advantage of access to government resources. Thus, the rural background of bureaucrats at local level tended to be identified with the interests of the green revolution entrepreneur. Probably it is this identification of the bureaucrats with the large farmers that made the small and poor farmers distrustful of government programmes.

In addition, the focus on technology was largely acceptable to both the bureaucrats and the political leadership in our study. Whatever differences that arose emanated from the problems of distribution of scarce inputs. There was keen competition among larger farmers to gain access to as much of fertilizers, irrigation facilities etc., and probably alignment with the local level bureaucrats was a necessary ingredient of the process. Thus, the political administrative conflicts were confined to a limited area. The general policy found support among both groups.

Further, the scope of bureaucratic action is influenced by the overall developmental framework in which green revolution was being implemented. The major assumption of agricultural development in India has been to rely on owner cultivated farms to produce more and provide surplus to the consumption stream. Policies of land reform were directed to provide for owner cultivation and more often than not, it is this component of land reform laws that has been implemented successfully and not the land ceiling component. Green revolution policies have been also more successfully adopted in those areas where tenancy problems have more or less been resolved. Owner cultivation is the norm in

Karnal and this group of farmers became the focus of new agricultural strategy.

In this framework, then, the role of bureaucracy was to promote greater production on these farms and to take the new technology to these farmers. The landless, the marginal or the small farmers were not the entrepreneurs whom the new agricultural strategy was looking for. What was being gained in agricultural production may have been lost on rural development but this was not part of the bureaucratic calculus.

In addition the market forces were not sought to be tampered with, on the other hand they were sought to be strengthened. In case of Brahm Prakash the interpenetration of credit, labour and product market determined his economic behaviour. The issue here is not whether the moneylender is exploitative or not but that the green revolution has led to interfaces of such types that such farmers have been left to the mercy of market forces. Institutional or bureaucratic intervention in this regard was not envisaged. On the other hand, the credit policies have actually strengthened the growing inability of such agricultural groups to participate in green revolution.

What this discussion points out is that the effectiveness and role of bureaucracy depends not only on the context of but also the content in which a policy is formulated and implemented. To the extent public policy intends to provide goods collectively without distinction among groups bureaucracy may be able to play an effective role. But if distinctive distribution is intended, then there may be serious limitations to bureaucratic role. Thus, if fertilizer was not in short supply and was to be distributed on first come basis, bureaucracy may have the capacity to perform this task effectively.

Similarly, if the intention of a policy is to bring about far reaching social changes through land reform or equitable distribution of agricultural inputs then bureaucracy may face difficulties. In such cases the bureaucracy is up against strongly entrenched vested interests wielding political power even if it is assumed that bureaucracy has the required "will" to take up cudgels against them.

Clearly then, the role of bureaucracy is tied to the groups that are propelling a particular type of development. As pointed out earlier, its functioning is not dependent on its internal competence alone but is determined to a great extent by the policies, their logic and intentions. What is important is to understand the social and political processes that give rise to a particular policy and demand its implementation in a particular way. It is this improved understanding that can provide a better perspective on the role and place of bureaucracy in policy implementation.

BIBLIOGRAPHY

Abrahamson, B. : *Bureaucracy or Participation : The Logic of Organization*, London, Sage Publications, 1977.

Appleby, Paul A. : *Public Administration in India : Report of a Survey*, New Delhi, Government of India, 1953.

Appu, P.S. : Tenancy Reform in India, *Economic and Political Weekly*, X, 33, 34 and 35, 1339-75, 1975.

Bhalla, G.S., and Alagh, Y.K. : *Performance of Indian Agriculture : A Districtwise Study*, Delhi, Sterling Publications, 1979.

Bhambhri, C.P. : Official—Non-Official Relations in Panchayati Raj, *Indian Journal of Political Science*, XVIII, 3, 1967.

Bhambhri, C.P. : *Administrators in a Changing Society*, Delhi, National Publishing House, 1972,

Bjorkman, J.W. : *The Politics of Administrative Alienation : Relations among Civil Servants and Political Leaders in Rural India : Develoment Programmes*, Mimeo, Doctoral Dissertation, 1976.

Braibanti, R. : Reflections on Bureaucratic Reform in India in R. Braibanti and J.J. Spengler (ed.), *Administration and Economic Development in India*, Durham, Duke University Press, 1963.

Braibanti, R. : *Asian Bureaucratic System Emergent from the British Imperial Tradition*, Durham, Duke University Press, 1966.

Chaturvedi, H.R. : *Bureaucracy and the Local Community : Dynamics of Rural Development*, Bombay, Allied, 1977.

Chaturvedi, T.N. : Tensions in Panchayati Raj : Relations Between Officials and Non-officials, *Economic Weekly*, 15 (22), 1964.

Dube, S.C. : *India's Villages*, London, Routledge and Kegan Paul. 1956.

Francine Frankel : India's New Strategy of Agricultural Development, *Journal of Asian Studies*, XVIII, 4, 1969.

Gaikwad, V.R. : *Panchayati Raj and Bureaucracy : A Study of the Relationship Patterns*, Hyderabad, National Institute of Community Development, 1969.

Government of Haryana : *Statistical Abstracts in Haryana*, Chandigarh, Yearly Series.

Government of India (Mehta Committee) : *Report of the Team for the Study of Community Projects and National Extension Service*, Delhi, 1957.

Government of India : *Modernizing Indian Agriculture Report on the Intensive Agricultural Development Programme*, 1960-68, Vol. I, Delhi, 1969.

Government of India : *Draft Fifth Five Year Plan*, New Delhi, Planning Commission, 1973.

Government of India : *Report of the Task Force on Agrarian Relations*, Delhi, Planning Commission, 1973.

Government of India : *Draft Five Year Plan*, 1973-83, New Delhi, 1978.

Gupta, Anil K. : A Note on Internal Resource Management in Arid Regions : Small Farmers—Credit Restraints : A Paradigm, *Agricultural Systems*, England, Vol. 6, No. 4 (Forthcoming), 1981.

Hunter, Guy : *Modernizing Peasant Societies*, London, Oxford University Press, 1969.

Huque, *et al* : *Theory of Rural Development*, Bangkok, 1975.

International Labour Organization : *Poverty and Landless in Rural Asia*, Geneva, 1977.

Khan, Waheeduddin, and Tripathy, R.N. : Cooperatives and Weaker Sections in Waheeduddin Khan (ed.), *Papers and Proceedings of the Workshop-cum-Seminar on Rural Institutions and Agricultural Development*, Hyderabad, National Institute of Community Development, 1972.

Kothari, S., and Ramashray Roy : *Relations Between Politicians and Administrators at the District Level*, New Delhi, Indian Institute of Public Administration, 1969.

Kothari, Rajni : *Politics in India*, Boston, Little Brown, 1970.

La Palombara, J. (ed.) : *Bureaucracy and Political Development*, Princeton, Princeton University Press, 1963.

Lipton, M. : India's Agricultural Performance : Achievements, Distortions and Ideologies in *Comparative Experience of Agricultural Development in Developing Countries of Asia and Far East Since World War II*, Bombay, Indian Society of Agricultural Economics, 1972.

Mathur, Kuldeep : A Committed Bureaucracy for India : Notes Towards Evolving a Model, *Political Science Review*, January-June 1971.

Mathur, Kuldeep : *Bureaucratic Response to Development*, Delhi, National Publications, 1972a.

Mathur, Kuldeep : *Sources of Indian Bureaucratic Behaviour : Organizational Environment and Political Processes in Rajasthan*, Jaipur, HCM State Institute of Public Administration, 1972b.

Mathur, Kuldeep : Conflict or Cooperation Administrators and Politicians in a Crisis Situation, *Indian Journal of Public Administration*, XX, 4, October-December, 1974, 835-45.

Mathur, Kuldeep : Administrative Institutions, Political Capacity and India's Strategy for Rural Development in Inayatullah (ed.), *Approaches to Rural Development*, Kualalumpur, APDAC, 1979.

Mathur, Kuldeep : Policies and Implementation of Land Reforms in India in Inayatullah (ed.), *Land Reform : Some Asian Experiences*, Kualalumpur, an APDAC, Publication, 1980.

Narain, Iqbal : Administrative Challenges in Panchayati Raj, *Indian Journal of Public Administration*, XII, 3, 1966.

Nehru, Jawahar Lal : *An Autobiography*, London, Oxford University Press, 1953.

Nicholson, Norman K. : *Panchayati Raj, Rural Development and the Political Economy of Village India*, Cornell, Cornell University, 1973.

Pai Panandikar and S.S. Kshirsagar : *Bureaucracy and Development Administration*, New Delhi, Centre for Policy Research, 1978.

Planning Commission : *Report of the Steering Group on Fifth Five Year Plan Relating to Agriculture, Irrigation and Allied Sector*, Delhi (Mimeo), 1973.

Programme Evaluation Organization, (PEO) : *Report on Evaluation Study of Small Farmers, Marginal Farmers and Agricultural Labourers*, New Delhi, Planning Commission, Government of India, 1979.

Punjab National Bank : *District Credit Plan, Karnal*, New Delhi, 1977.

Punjab National Bank : *District Credit Plan 1980-82, Karnal*, Delhi, 1980.

Reserve Bank of India : *Report of the All India Rural Credit Review Committee*, Bombay, 1969.

Savage, Peter : Dismantling the Administrative State Paradigm Reformulation in Public Administration, *Political Studies*, XXII, 1974, 147-57.

Sharma, J.P., and D.S. Mehra : *Rural Development : Choice of Programme and Strategy*, New Delhi, IIPA, Mimeo, 1978.

Singh, K.K. and Ali Ashraf : *Bureaucracy, Leadership and Development*, New Delhi, Report Submitted to ICSSR, (Mimeo), 1971.

Taylor, C.C., *et al* : *India's Roots of Democracy*, New York, Praeger, 1966.

Viet, Lawrence A. : *India's Second Revolution the Dimensions of Development*, New York, McGraw Hill, 1976.

Index